Digital Communications

Digital Communications

Course and Exercises with Solutions

Pierre Jarry
Jacques N. Beneat

First published 2015 in Great Britain and the United States by ISTE Press Ltd and Elsevier Ltd

ISTE Press Ltd
27-37 St George's Road
London SW19 4EU
UK

Elsevier Ltd
The Boulevard, Langford Lane
Kidlington, Oxford, OX5 1GB
UK

www.iste.co.uk

www.elsevier.com

Notices

Knowledge and best practice in this field are constantly changing. As new research and experience broaden our understanding, changes in research methods, professional practices, or medical treatment may become necessary.

Practitioners and researchers must always rely on their own experience and knowledge in evaluating and using any information, methods, compounds, or experiments described herein. In using such information or methods they should be mindful of their own safety and the safety of others, including parties for whom they have a professional responsibility.

To the fullest extent of the law, neither the Publisher nor the authors, contributors, or editors, assume any liability for any injury and/or damage to persons or property as a matter of products liability, negligence or otherwise, or from any use or operation of any methods, products, instructions, or ideas contained in the material herein.

For information on all our publications visit our website at http://store.elsevier.com/

British Library Cataloguing-in-Publication Data
A CIP record for this book is available from the British Library
Library of Congress Cataloging in Publication Data
A catalog record for this book is available from the Library of Congress
ISBN 978-1-78548-037-9

Printed and bound in the UK and US

Contents

Preface

Digital communications plays an important role in numerical transmission systems because of the proliferation of radio beams, satellites, optic fibers, radar and mobile wireless systems.

This book provides *fundamentals and basic design techniques* of digital communications and that is why the manuscript is *brief and quite simple*.

It has grown out of the authors' own teaching and as such has a unity of methodology and style, essential for a smooth reading.

The book is intended for engineers and for advanced graduate students.

Each of the 13 chapters provides a complete analysis of the structures used for emission or reception technology. We hope that this will provide the students with a set of approaches that he/she could use for current and future digital circuits designs.

We also emphasize the practical nature of the subject by summarizing the design steps and giving examples of exercises so that digital communications students can have an appreciation of each circuit. This approach, we believe, has produced a coherent, practical and real-life treatment of the subject.

The book is therefore theoretical but also experimental. The exercises represent more than 25% of the book.

The book is divided into four parts and 13 chapters. Then we have to give the different systems of telecommunication (radio beams, satellites, optic fibers, radar and mobile telephones) and also the principles of the baseband transmission, the digital channels and the different digital communications.

In Part 1, we give the definition of the digital communications and the different channels that propagate these digital elements.

In Chapter 1, we focus on the definitions and on the quality of the digital communications. Chapter 2 is entirely devoted to the digital radio beams, satellites, optic fibers and Radars.

After we have given the used frequencies, we describe the equipment (antennas, microwave oscillators, microwave amplifiers, filters, etc.).

We show that it is possible to transmit to a long distance by studying losses and dispersion.

In Part 2, we define the different sampling methods, the analog-to-digital conversion, the errors due to these sampling and the quantifications. In Chapter 3, the different ways of pulse-code modulation and line coding are described: RZ (return to zero) and NRZ (no return to zero). The channel is described in Chapter 4 and we give the different S/N ratios, the possibilities to regenerate the signal, the detection in the case of a loss support and the eye pattern. We also give the method of equalization of the transfer function of these supports. Chapter 5 is entirely devoted to the three criteria of Nyquist. We conclude by giving the error probability of the whole line. This chapter is the only chapter that is a little theoretical. In Chapter 6, we make analog-to-digital conversion by sampling and uniform and non-uniform quantization of the signal. We define the compression that is necessary to coding. We give the different noises in the cases of a sinusoidal signal and of a white noise. After compression or before extension, we introduce codes as series/parallel or DELTA.

Part 3 is entirely devoted to digital communications on the carriers' frequencies. In Chapter 7, we give the amplitude modulation and the phase shift keying (PSK) of the carrier. Comparison of the two modulations shows that only the PSK is coherent. Chapter 8 is devoted to the differential coding of orders 2 and 4. In Chapter 9, we give the methods to demodulate (detect) the signals at the reception. Having a PSK-2 and a PSK-4 at the emission, we give the classical method to recover the signal at reception using a coherent demodulation. It is also shown that it is possible to use differential demodulations (orders 2 and 4). In Chapter 10, we give the quality of the different modulations.

In Part 4, we have various exercises and problems. In Chapter 11 first we deal with the different supports as power links or antenna. We also give a simple method for designing duplexers and multiplexers. Chapter 12 gives problems corresponding to the different modulations and demodulations without noise (system of amplitude modulation, system of coherent amplitude demodulation and the quadrature system).

However, in Chapter 13, the noise is introduced into problems concerning the modulations and demodulations (noisy demodulation using triangular filters, detection in the presence of noise and analog-to-digital conversion).

These courses and problems with solutions were given during masters level engineering courses.

Starting from these fundamentals, the students can deepen their knowledge by using complementary and valuable books of courses and exercises as those from H. Taub and D. L. Schilling, F. G. Stremler, L. E. Larson, R. E. Ziemer and W. H. Tranter, L. W. Couch II, G. Maral and M. Bousquet, M. Schwartz, or A. B. Carlson [STR 82, COU 83, ZIE 85, TAU 86, JAR 14, JAR 15, CAR 87, SCH 87, LAR 97, MAR 98].

Professor Pierre JARRY (France)
Professor Jacques N. BENEAT (USA)
September 2015

Acknowledgments

These courses were provided in France at the universities of Limoges, Brest and Bordeaux.

Pierre Jarry wishes to thank his colleagues at different universities and schools of engineers.

He would also like to express his deep appreciation to his wife and his son for their support.

Jacques N. Beneat is very grateful to Norwich University in USA, a place conducive for trying and succeeding in new endeavors.

Finally, the authors express their sincere appreciation for all the staff at ISTE involved in this project for their professionalism and outstanding effort.

Part 1

Telecommunication Systems

Digital Communications

1.1. Introduction

The information has to be transmitted in the form of binary digit (bit). To do this, the analogical signal is converted into a (digital) numerical signal by sampling the signal (Figure 1.1).

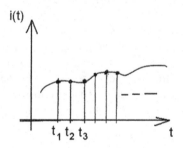

Figure 1.1. *Sampling the signal*

We suppose that we can directly transmit the samples of the signal, but this method is almost perturbed by noise. Then we quantify these samples and code, which gives a binary description.

For example, if we consider a binary number of 8 elements:

– quantification will have $2^8 = 256$ levels;

– all the levels are performed with a binary number (binary digit or a "bit") of 8 elements.

We must sample the signal rapidly and if the sampling frequency is 8 kHz, we have 8,000 samples/second and the frequency of a "bit" will be 8 × 8,000 bits/second.

1.2. Qualities of the digital communications

In the (digital) numerical domain, the transmission is characterized by the "binary error rate" or BER, which means that one binary element ("1" or "0") is recognized or not recognized at the output. Then if BER = 10^6, there is one false binary element for 10^6 binary elements transmitted, and the probability error is $\mathcal{P} = 10^{-6}$.

In the analogical domain, we have to transmit the binary elements "1" or "0" on the channel, and it is characterized by the "signal-to-noise" (S/N) ratio. If the channel (radio beams, satellites, optical fibers, radar, etc.) introduces perturbations in the form of distortions and noise, then we need a high S/N ratio (7–8 dB).

1.3. Advantages of digital communications

In terms of advantages, we can say that digital communications have a good resistance to degradations because of the presence of repeaters that also regenerate the signal at finite distances. Coding that can detect and correct the errors also exists. They have a property of transparency: the same channel can transmit a lot of information (speech, images, data, etc.). They are also economically advantageous because the extremities' equipment is cheaper than that in the analogical domain (filters, multiplexers, etc.).

But the numerical transmissions are bulky for their passband (the numerical signal has large spectrum).

1.4. Definitions

At the emission, the baseband (B&B) transmission is the *processing* of the information in the frequency band ranging from 0 to f_C Hz, where f_C is the cutoff of the *signal processing*.

In the B&B transmission, the coding is termed as *line coding*.

But the information is transmitted on f_p, which is the *carrier frequency*.

The transmission by the carrier is called *modulation*.

At the reception, the signal has to be *demodulated* on the carrier frequency and *deprocessed* in the B&B domain.

1.5. Diagram of digital communications

A system of digital transmission takes the form shown in Figure 1.2.

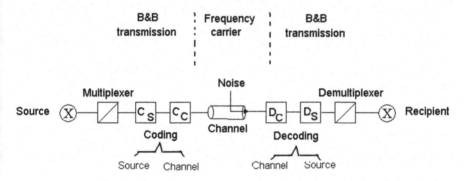

Figure 1.2. *Diagram of the digital communication*

The signals travel from the source to the recipient. After multiplexing, they are first coded and then carried by the channel. They are then decoded and at the end are demultiplexed to go to the output.

We have to consider two kinds of coding: the source coding and the channel coding.

Coding and multiplexing (or demultiplexing) are performed in the B&B domain whereas transmission on the channel (radio beams, satellites, optic fibers, radar, etc.) is performed by a frequency carrier.

2

Supports of Digital Communications

2.1. Introduction

The transmission's supports are the radio beams, the satellites, the optic fibers, and the radar. We give some quick information on these support elements.

2.2. The radio beams

The radio beams can transmit all kinds of signals such as voices (conversations), television programs, or data. We consider only one jump that is about 50 km (Figure 2.1).

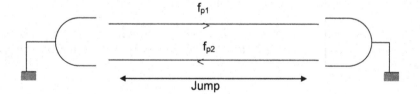

Figure 2.1. *A jump between two antennas*

The two ways of transmission (go and return) are carried by two frequencies f_{P1} and f_{P2}, which are called the modulation frequencies. We also consider digital radio beams and the carrier is a "phase shift keying" (PSK). The antennas are common and the transmission needs several jumps (Figure 2.2).

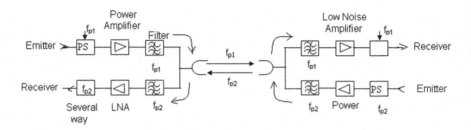

Figure 2.2. *Diagram of a radio beam*

2.2.1. *The carrier frequencies*

The carrier frequencies are in three microwave domains:

– ultrahigh frequencies ranging from 300 MHz to 3 GHz, which gives a wavelength ranging from 1 m to 10 cm;

– superhigh frequencies ranging from 3 to 30 GHz, with a wavelength ranging from 10 to 1 cm;

– extremely high frequencies ranging from 30 to 300 GHz, with a wavelength ranging from 1 cm to 1 mm.

The bands included in the frequency range from 1.7 to 12 GHz are reserved for the analog radio beams and we will not discuss them. However, in order to have more capacity, the radio beams are now no longer included in the bands ranging from 12 to 30 GHz, and are reserved for the digital radio beams.

The choice is also given by the parasite coupling between the antennas situated on the same support (Figure 2.3) and by the selectivity of the emitters and the receivers.

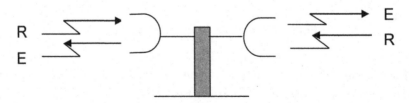

Figure 2.3. *Parasite couplings*

2.2.2. The antennas

In general, antennas are made with parabolic reflectors (Figure 2.4). Their focus is occupied by the "source" so that the outputs give parallel rays. In the second case (Figure 2.4), the focus is outside scattering ("flux"?) and there is no shadow zone.

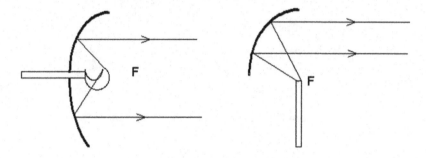

Figure 2.4. *Parabolic antennas*

The gain g of is given by:

$$g = r\frac{4\pi}{\lambda^2}A$$

where A is the geometric aperture, r is the output factor (approximately 0.5 and depends on the antenna form) and λ is the wavelength. Then g is varying as f^2, where f is the considered frequency. The choice of the antenna depends on:

– the necessary gain;

– the directivity;

– the carrier frequencies;

– the available pace;

– the cost, etc.

2.2.3. The connections over the horizon

When it is not possible to install relay stations (geographic or politic reasons), we can use the scattering (diffusion) phenomena on the troposphere (Figure 2.5).

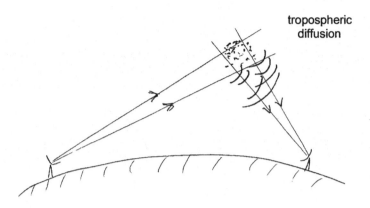

Figure 2.5. *Connections over the horizon*

The intensities of the secondary reflected waves are very less and the qualities of these connections are not very good. That is why we prefer the transmissions via satellites over long distances (more than 1000 km).

2.3. The satellites

Communication across seas and oceans has always been a technical fascination and a technological challenge. With this rapid historical, we can say that it is now possible:

– 1858: first transatlantic telegraph cable;

– 1901: wireless telegraph from Great Britain to Newfoundland (Terre-Neuve);

– 1927: telephonic transatlantic connection using short waves;

– 1957: telephonic transatlantic coaxial cable;

– 1962: first active satellite "Telstar I" and transatlantic transmission for the television;

– 1965: geostationary satellite "EarlyBird" or "Intelsat I", Intelsat is a cooperative society with more than 100 member countries;

– since 1965: the satellite missions are now classical and numerous.

Apart from telecommunications (which are for telephone, for television, for data, etc.), satellites can also be used for:

– military purposes;

– maritime navigation;

– meteorology;

– teledetection (natural resource).

2.3.1. The frequency domains

Some frequencies (or group of frequencies) are reserved for communications by satellites, which are as follows:

– 6–4 GHz;

– 14–11 GHz;

– 30–20 GHz.

For example, in the first case, $f_{pu} = 6\,\text{GHz}$ is the carrier frequency for the uplink way (from earth to satellite) and $f_{pd} = 4\,\text{GHz}$ is the carrier frequency for the downlink way (from satellite to earth), as shown in Figure 2.6. Then, it is necessary to transpose the carrier from $f_{pu} = 6\,\text{GHz}$ to $f_{pd} = 4\,\text{GHz}$ or from $f_{pd} = 4\,\text{GHz}$ to $f_{pu} = 6\,\text{GHz}$ (Figure 2.6).

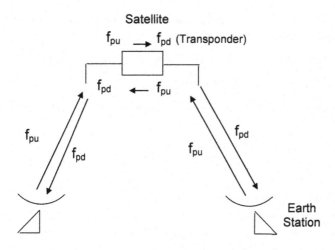

Figure 2.6. *Diagram of a satellite*

In all the cases, we have to make the choice by keeping in mind:

– the troposphere absorption;

– the atmospheric and cosmic perturbations.

Then the satellites are principally made by repeaters that amplify and transpose the signal.

2.3.2. The noises

Essentially three noise sources that perturb the signal exist. These are:

– the thermal noise, which is due to the input resistance of the reception amplifier;

– the internal noise of the receiver;

– the external noise of the antenna (cosmic, beam from the earth).

In all the cases, we measure the temperature of the noise.

2.3.3. Comparison with the radio beam

If we make a comparison with the radio beam we can see that there are advantages but is also inconvenient, for example:

– the emitting power of the satellite is limited;

– the antenna gain of the satellite is modest;

– but a distance of 2500 km needs 50 jumps by a radio beam and only 2 jumps by a satellite!

2.3.4. Diagram of a satellite connection

As an example we show the diagram of a satellite connection of the Intelsat IVA type in the band 6–4 GHz (Figure 2.7). We can see that a signal of 1.2 kW power goes up and returns to the earth via the satellite with a slight power of 1.2 pW. But the information has not been lost during the way.

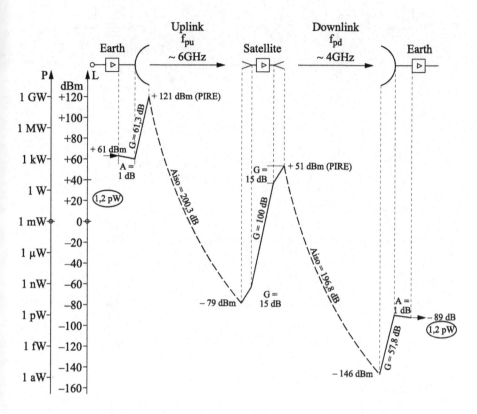

Figure 2.7. *Power diagram of a satellite connection with* $P_{dBm} = 10\log\left(P_{Watts}/10^{-3}\right)$

2.3.5. Transponders

The transponders (Figure 2.8) are necessary to:

– regenerate the signal in the noise;

– amplify this signal;

– transpose the signal from f_{pu} (uplink) to f_{pd} (downlink).

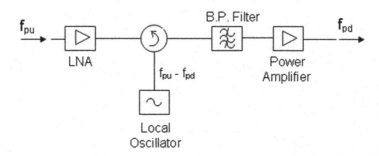

Figure 2.8. *Transposition of carrier frequency*

2.3.6. *Prospects*

Because of the growing number of frequencies and spatial traffic:

– we have to increase the capacities of the satellites (several hundreds of thousands of the ways);

– the band 14–11 GHz is now exhausted and we need to exploit the band 30–20 GHz;

– we need to exploit domestic satellites (inside the same country).

2.4. The optical fiber

Transmission systems on optical fiber are digital (Figure 2.9). Its functions are similar to those of a classical system.

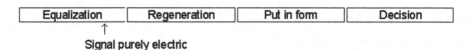

Figure 2.9. *The functions of a transmission system*

In the case of the optical fiber, the optic signal must pass through a photodiode to deliver an electric signal but with a shot noise (a Poisson noise that depends on the emitting signal).

This shot noise is in addition to the thermal noise (a Gaussian noise) of the receptor, which is a characteristic of the optical fiber transmissions (Figure 2.10).

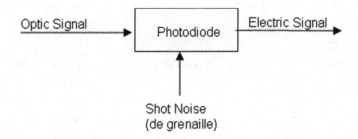

Figure 2.10. *The shot noise of the photodiode*

To familiarize ourselves with the digital transmissions on the optical fibers, we have to successively and rapidly give the different elements of the chain:

– the optical fiber as the support;

– the laser as generator;

– the photodetector as receptor.

2.4.1. *Propagation and structure of the optical fiber*

Resolution of Maxwell equations gives us electric and magnetic waves and then it is possible to get the propagation conditions. The profile of the optical fiber (Figure 2.11) comprises two mediums of different indices (n): the heart (which is the guide part or core) and the sheath (or cladding).

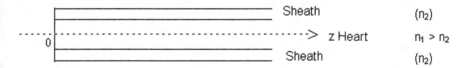

Figure 2.11. *Profile of the optical fiber*

We call this very smooth variation as the index profile (variation about 1%). The possible profiles of Figure 2.12 show three types of optical fibers:

– mono-mode profile with a jump of index (step-index fiber);

– multimode profile with a jump of index (also step-index fiber);

– multimode profile with a gradient of index (graded-index fiber).

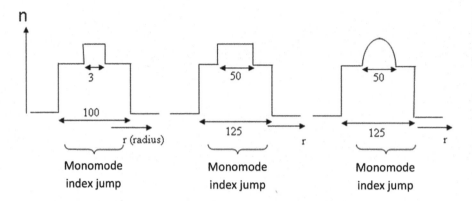

Figure 2.12. *Different profiles of the optical fibers*

2.4.1.1. Step-index fiber

Propagation is made by multiple reflections at the boundary of the two mediums (Figure 2.13).

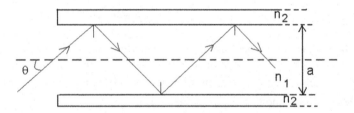

Figure 2.13. *Step-index fiber*

There is total reflection if the value of θ is less than a critical value θ_C, so that:

$$\sin \theta \le \sin \theta_C = \sqrt{n_1^2 - n_2^2} = NA$$

where NA is the numerical aperture (in general, the value is about 0.17).

2.4.1.2. Graded-index fiber

The guide is assimilated as a piling up of homogeneous layers with constant indices and with infinitesimal thickness (Figure 2.14). It is shown that oblique rays have greatest speed than those evolving parallel to the z-axis.

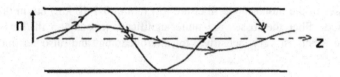

Figure 2.14. *Graded-index fiber*

2.4.1.3. *Mono-mode fiber*

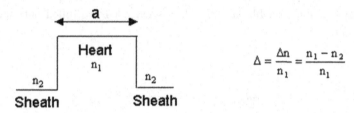

$$\Delta = \frac{\Delta n}{n_1} = \frac{n_1 - n_2}{n_1}$$

Figure 2.15. *Mono-mode fiber*

The mono-mode fiber has only one mode of propagation HE_{11}, which is a hybrid mode (Figure 2.15). Then the numerical aperture NA is:

$$NA = \sqrt{n_1^2 - n_2^2} = \sqrt{(n_1 + n_2)\, \Delta n} \approx \sqrt{2 n_1\, \Delta n}$$

We have just to adjust the parameters of the fiber to make it mono-mode fiber.

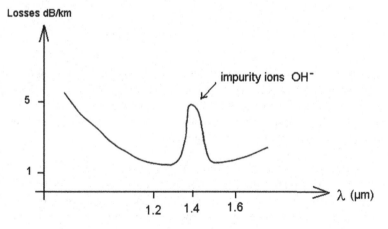

Figure 2.16. *Losses versus wavelength*

The losses of the mono-mode fiber are low (about 0.4 dB/km) but in the case of the multimode fiber the losses are quite significant (about 4–5 dB/km). But the mono-mode fiber is very sensitive to microbending and also to mechanical constraints.

We use wavelengths of approximately 0.8, 1.3 or 1.55 μm that correspond to regions with low losses (Figure 2.16).

2.4.2. Fabrication technology of the optical fiber

We use a double crucible and the fiber is extended and rolled onto the drum (Figure 2.17).

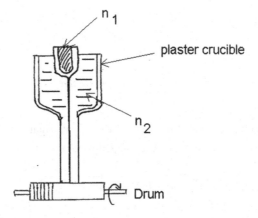

Figure 2.17. *Fabrication of a fiber*

2.4.3. Lasers

Optic waves are electromagnetic radiations with very short wavelengths ranging from 0.4 μm (purple) to 0.8 μm (red). The sources are monochromatic (they deliver only one frequency or a short band). They are also coherent, which is essential to transport pulses (the phase of the oscillators has to be well defined and invariable with time). The source has also to be integrated.

The semiconductor laser is the solution of all these problems, and this source emits light from a current. The optic power of a laser diode is from 10 to 25 mW. From Figure 2.18, it is clear that an important characteristic of this laser electroluminescent diode is the curve of the emitting optic power P in function of the polarization current I. We can see that the laser effect appears at a threshold I_T.

P(mW)

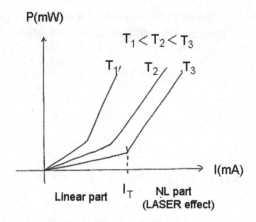

Figure 2.18. *Characteristic of a laser diode*

This characteristic permits us to obtain a digital modulation with the input as current *I* and the output as optic power *P(I)* for a given temperature T_i (Figure 2.19).

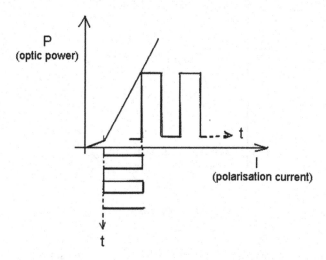

Figure 2.19. *Optic power versus current polarization*

2.4.4. *The photodetector*

A photodetector is also a PN junction that transforms the light (photons of the optic fiber) into an electric current. As shown in Figure 2.20, the photodiode is equivalent to a generator of current of the signal I_S in parallel to a noise current I_B, in parallel to the junction capacity C_j and to the load resistance R_L. The photodetection phenomenon is the inverse of the semiconductor laser effect: this is the optic absorption.

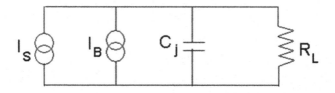

Figure 2.20. *Equivalent circuit of a photodiode*

We give absorption α in function of the wavelength λ for the classical semiconductor materials used in optic telecommunications (Figure 2.21).

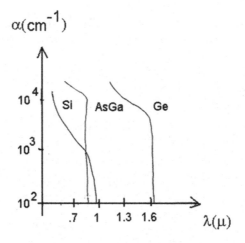

Figure 2.21. *Absorption in the cases of Si, AsGa and Ge*

None of the three materials (Si, AsGa and Ge) cover the useful wavelength band that ranges from 0.7 to 1.6 µm.

We should also be aware that a PIN photodiode is a fast and sensitive diode and is used in the case of short distances. It is made up of an intrinsic material I with high resistivity enclosed by materials P and N^+, both having low resistivity.

In the case of longer distances, we use an avalanche diode that has a numerical high speed. Then we have to associate an amplifier with the reception chain.

2.4.5. *Transmission system of optical fiber*

The diagram of the optic fiber transmissions is shown in Figure 2.22. Then we have to remember that there are two types of noise at the reception:

– a thermal noise (random and Gaussian that is independent of the emitting signal, the same as that in radio beam and satellite);

– a shot noise (Poissonian that depends on the emitting signal).

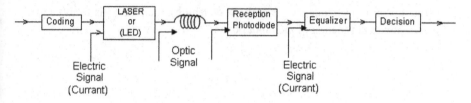

Figure 2.22. *Diagram of the optic fiber transmissions*

2.5. The radar

Radar stands for radio detection and ranging. The signal is transmitted, then partially reflected by an object (the target) and detected. We can determine the distance of the object, its speed, its form, etc.

2.5.1. *Applications of radar*

Radar has applications in the civil, military, and scientific domains.

Civil applications concern planes, boats, climate, altimeters, landing, speed measurement (policy), etc.

Military applications concern air and sea navigation, detection (such as planes and missiles), missile and rocket guidance, identification, etc.

The scientific domains include astronomy, cartography, measuring of distances, natural energy resources, etc.

2.5.2. *The radar equation: the Friis formula*

Let us consider two stations: one emitting E and one at the reception R (Figure 2.23).

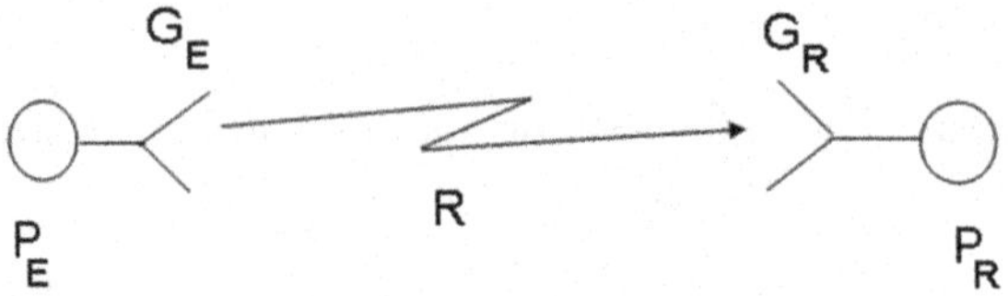

$$G_E \qquad\qquad\qquad G_R$$

R

$$P_E \qquad\qquad\qquad\qquad\qquad P_R$$

Figure 2.23. *Two stations*

If f is the emitting frequency (or λ the wavelength), then the Friis formula gives the output power in function of the input power:

$$\frac{P_R}{P_E} = \frac{G_R\,G_E\,\lambda^2}{(4\pi R)^2}$$

where R is the distance between the two stations, P_E is the emitting power, P_R is the receiving power, G_E is the antenna gain on the emission side and G_R is the antenna gain on the reception side.

The radar equation is shown in Figure 2.24.

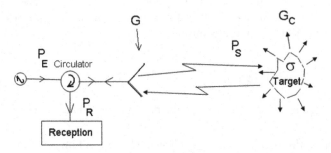

Figure 2.24. *The radar formula*

From the source to the target we have:

$$P_S = \frac{G\,G_C\,\lambda^2}{(4\pi R)^2}\,P_E$$

And also from the target to the reception:

$$P_R = \frac{G\,G_C\,\lambda^2}{(4\pi R)^2}\,P_S$$

Then, we move from the source (emitting) to the reception going through the circulator:

$$P_R = \frac{G^2 \lambda^2}{(4\pi)^3 R^4} \left(\frac{G_C^2 \lambda^2}{4\pi} \right) P_E$$

This means with the radar cross section $(RCS = \sigma = G_C^2 \lambda^2 / 4\pi)$:

$$P_R = \frac{G^2 \lambda^2}{(4\pi)^3 R^4} \sigma P_E$$

This is the radar equation, which means that the reception power varies as R^{-4}. Then we must use very low noise detector for long distances.

2.5.3. Radar cross section

This surface gives the report of the scattering power on the incident power of the target. We give some typical values of $\sigma(m^2)$ (Figure 2.25).

target	$\sigma(m^2)$
bird	0.01
missile	0.5
individual	1
small plane	1 – 2
bicycle	2
small ship	2
fighter plane	3 – 8
bombardier	30 – 40
cargo plane	100
wagon	200

Figure 2.25. *Typical values of RCS*

RCS of complex targets, such as planes or ships, varies rapidly with the frequency. Then it is useful to minimize RCS with absorbing material.

Part 2

Baseband Digital Communications

Baseband Communications

3.1. Sampling the signals

We have to transmit the signal $s(t)$ at the discreet times t_i, and $s(t_i)$ are the samples. If these samples $s(t_i)$ are quite numerous, then it is possible to recover the signal $s(t)$. This is a consequence of Shannon's theorem.

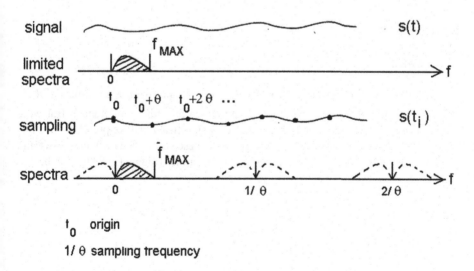

Figure 3.1. *Sampling of a signal*

From Figure 3.1 we can see that, if the spectrum of $s(t)$ is limited to f_{MAX}, it is possible to restitute $s(t)$ from the values of the samples $s(t_i)$ at the times $t_0 + i\theta$, provided the spectra at $0, 1/\theta, 2/\theta, ..., i/\theta, ...$ do not overlap themselves. Then we must have:

$$\frac{1}{\theta} \geq 2f_{MAX}$$

Then we can recover the signal $s(t)$ by a simple filtering operation.

The spectrum of $s(t)$ has to be limited and is obtained by a low-pass filtering (0 to f_{MAX}).

As first practical example, we consider the signal of the telephone. To have a good auditory quality, we must consider the band range of 300–3,400 Hz, which corresponds to a value of $\theta = 125\ \mu s$ or a sampling of 8 kHz.

A second practical example is the case of television in which we must consider the band range of 0–6 MHz and a sampling of $\theta = 83\ ns$ or 12 MHz.

3.2. Coding the samples

3.2.1. *Basis*

Now we have the samples of Figure 3.2 and we can imagine transmitting $s(t_i)$ directly onto the channel (radio beams, satellites, optic fibers, or Radar). But with the problems of the additive noises, it is not very significant and we prefer to code the samples and transmit a number that corresponds to these codes. The method is powerful because it is always possible to regenerate the transmitted signal (in fact a number).

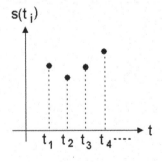

Figure 3.2. *The samples $s(t_i)$*

The signal to be coded must be less than V_{MAX} and this quantity is called the *"dynamic"* of the encoder. The sample is assigned a sign $+$ or $-$ and an absolute value V such that $V \le V_{MAX}$.

The basis is 2 because the binary operations are easily realizable with electronic circuits.

The interval $[0, V_{MAX}]$ is divided into 2^n minus intervals $[V_i, V_{i+1}]$, if n is the order of the encoder.

$$
\left\{
\begin{array}{l}
V_0 = 0 \\[2mm]
V_1 = \dfrac{V_{MAX}}{2^n} \\[4mm]
V_2 = \dfrac{V_{MAX}}{2^n} \cdot 2 \\[2mm]
\vdots \\[1mm]
V_p = \dfrac{V_{MAX}}{2^n} \cdot 2^p \\[2mm]
\vdots \\[1mm]
V_{MAX} = \dfrac{V_{MAX}}{2^n} \cdot 2^n
\end{array}
\right.
$$

To give the value of a sample, we pose ourselves the following question: in which interval is V placed?

$V_{MAX}/2^n$ is the unity of the encoder, where n is the moment of the encoder and if $V_i \le V \le V_{i+1}$ then V is taken as V_i.

Then the value attributed to the sample is:

$$\left(0, 1, ..., 2^p, ..., 2^{n-1}\right)$$

These numbers can be represented in the base 2 with n numbers (Figure 3.3). For example, in the simple case $n = 3$, we have the next.

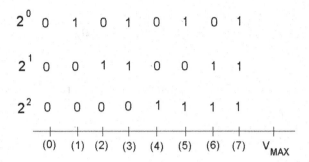

Figure 3.3. *The numbers in the base 2 with n = 3*

3.2.2. Binary representation

The binary representation uses only the binary digit (bit): 0 and 1. There is a supplementary bit that gives the sign of the sample. Then the sample is translated by a word of $n + 1$ bits.

If T is the length of the emission of the bit 0 or 1, then $1/T$ is the modulation rapidity (in Bauds).

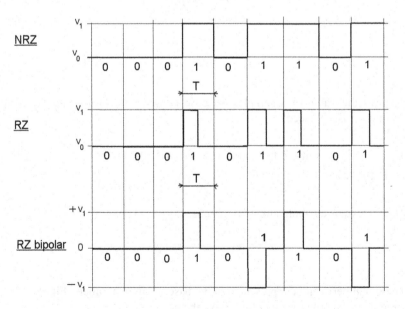

Figure 3.4. *An example of the codes NRZ, RZ and RZ bipolar*

For the no return to zero (NRZ)   code, tension V_0 corresponds to the digit 0 whereas tension V_1 corresponds to 1 (Figure 3.4). For the return to zero (RZ) code also, tension V_0 corresponds to the digit 0 whereas tensions V_1 and V_0 correspond to 1. In the case of the *RZ* bipolar code, tension $V_0 = 0$ corresponds to the digit 0 whereas tensions V_1 and $-V_1$ correspond to 1. The sign of V_1 is chosen to alternate with the previous tension V_1.

The bit rate is the quantity of bits transmitted per second (bit/s). It is different from the modulation rapidity!

3.3. Time multiplexing

In the numerical domain, we consider time multiplexing. As an example, we have two pulses *A* and *B* with the same modulation rapidity $1/T$ and coded in NRZ (Figure 3.5).

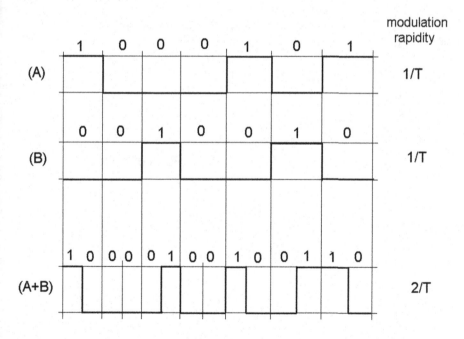

Figure 3.5. *An example of NRZ time multiplexing (duplexing)*

This is true in the case where the two pulses (A and B) are perfectly synchronous. In fact, the pulses have a slight phase difference because they are controlled by the same clock ($T_A \neq T_B$). In this case, we add a supplementary bit or pulse to reduce this difference of phase to zero. This is called a stuffing pulse.

In general, we must remember that the modulation rapidity of the multiplexer obtained from n numerical trains is n times the modulation rapidity of each of the components.

3.4. Spectrum of the different binary representations

We give the spectral density of power in function of the frequency $S_x(f)$. This quantity is used to determine the perturbations of one channel to another. This is the problem of the crosstalk.

3.4.1. NRZ code

The amplitude $-V$ represents the level "0" and $+V$ represents the level "1" during the time T. The spectrum is continuous (Figure 3.6):

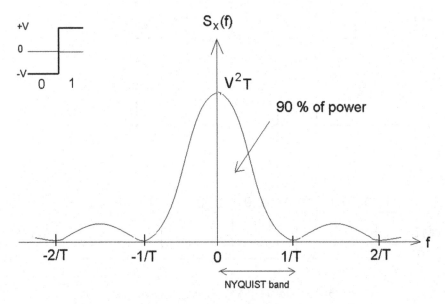

Figure 3.6. *Continuous spectrum of the NRZ code*

$$S_x(f) = V^2 T \left(\frac{\sin \pi f T}{\pi f T} \right)^2$$

From the continuity of the spectrum of the NRZ code, it is difficult to recover the frequency clock.

3.4.2. RZ code

The spectrum of the RZ code presents a continuous and a discontinuous part (Figure 3.7).

$$S_x(f) = \frac{V^2 T}{4} \left(\frac{\sin \pi f T / 2}{\pi f T / 2} \right)^2 + \frac{V^2}{4} \delta(f) + \frac{V^2}{\pi^2} \sum_{n=-\infty}^{+\infty} \frac{1}{(2n+1)^2} \delta \left(f - \frac{2n+1}{T} \right)$$

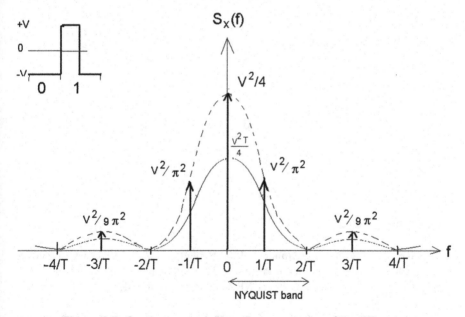

Figure 3.7. *Continuous and discontinuous spectra of the RZ code*

If a part of the spectrum of the RZ code is non-continuous, then it is not difficult to recover the frequency clock. This is an important property but the band is (two times!) broader than that in the case of the NRZ code.

3.4.3. *RZ bipolar code*

The spectrum of the RZ bipolar code presents only a continuous part:

$$S_x(f) = \frac{V^2 T}{4} \left(\frac{\sin \pi f T / 2}{\pi f T / 2} \right)^2 \sin^2 \pi f T$$

It has no power at the frequency zero and then it is adapted in the case of transmissions that require transformers (Figure 3.8).

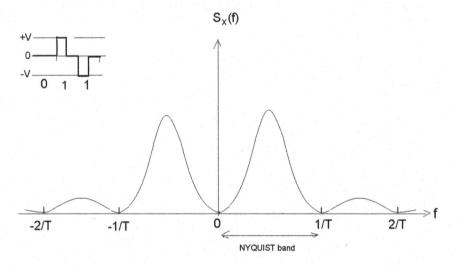

Figure 3.8. *Continuous spectrum of the RZ bipolar code*

3.5. Frequency of the clock

The RZ code presents a component at the frequency $1/T$ and it is easy to recover the frequency of the signal clock by filtering. This can be done without difficulty but the spectrum of RZ is twice that of the NRZ and RZ bipolar codes

In the case of the NRZ and RZ bipolar codes, it is possible to recover the frequency of the signal clock using a nonlinear processing and a filter.

4

The Channel

4.1. Regeneration of the signal: general diagram

This section describes different channels such as radio beams, satellites, optic fibers and radars.

L_0 is the distance of regeneration between two repeaters that also regenerate the signal (R.R. is a repeater regenerator). In the case of telephone, we have $L_0 = 1,830$ m.

The transmission can be made in two ways: 1 is direct and 2 is back (Figure 4.1).

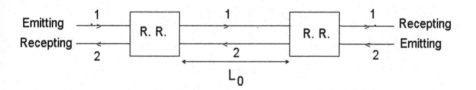

Figure 4.1. *Repeaters - Regenerators*

We emit a sequence of "0" and "1" in a given code, which is degraded by the noise and the support (Figure 4.2).

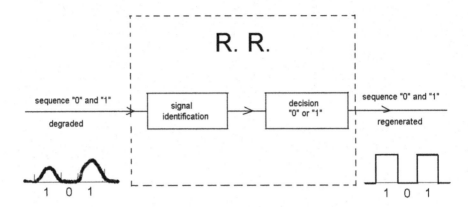

Figure 4.2. *Regeneration of the signal*

After identification and decision, the regenerated sequence of "0" and "1" at the output must be identical to that at the input (Figure 4.2).

It is in general not true because there is the influence:

– of the noise;

– of the distortion of the support.

We will examine these two influences: first the noise and then the support.

4.2. Detection of a noisy digital signal

We will consider in order:

– the cases in which the support is considered to be perfect (then the noise is additive);

– the defaults of the channel.

4.2.1. *Perfect support*

Without the loss of generality, we consider the NRZ signals. At the input of the R.R., we have the signal:

$$x(t) = \sum_k a_k s(t - kT) + n(t)$$

where $s(t)$ is the elementary pulse of T duration (Figure 4.3).

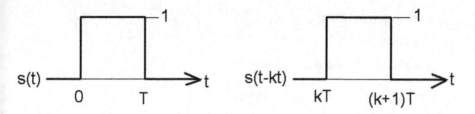

Figure 4.3. *Elementary pulse of T duration*

In the case of the NRZ code, the values of the coefficient a_k can be only:

$$a_k = 0 \text{ or } 1$$

where $n(t)$ is the thermal noise created by the input of the receptor. This is a white Gaussian or a quasi-white Gaussian noise (with a constant or quasi-constant power spectral density).

The basic signal $s(t)$ is of T duration and receives the information during the time interval:

$$\left[kT, (k+1)T \right]$$

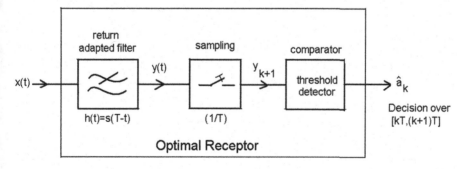

Figure 4.4. *Optimal receptor*

The decision is taken symbol by symbol, and we show that the optimal receptor is a cascade of (Figure 4.4):

– a return adapted filter;

– a sampler;

– a comparator (threshold detector).

If $\hat{a}_k = a_k$, then the decision is correct.

But if $\hat{a}_k \neq a_k$, the decision is incorrect (i.e. there is an error).

Consider $\mathcal{P}_e$ as the error probability, so that $\hat{a}_k \neq a_k$ and $\mathcal{P}_d = 1 - \mathcal{P}_e$ the correct decision.

After going through the return adapted filter, we have:

$$y(t) = x(t) * h(t)$$
$$y(t) = \sum_k a_k s(t - kT) * h(t) + n(t) * h(t)$$

Using:

$$\left[s(t) * \delta(t - kT) \right] * h(t)$$
$$= \left[s(t) * h(t) \right] * \delta(t - kT)$$
$$= u(t) * \delta(t - kT)$$
$$= u(t - kT)$$

we have:

$$\boxed{y(t) = \sum_k a_k u(t - kT) + n(t) * h(t)}$$

With:

$$u(t) = s(t) * h(t) \quad \text{and} \quad h(t) = s(T - t)$$

then:

$$u(t) = \int_{-\infty}^{+\infty} s(\tau)h(t-\tau)d\tau = \int_{-\infty}^{+\infty} s(\tau)s(T-t+\tau)d\tau$$

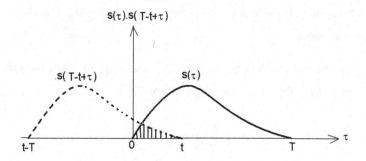

Figure 4.5. *Variations of* $s(\tau)$ *and* $s(T-t+\tau)$

We can see that $u(t)$ is maximum at the sampling moment $t = T$.

Then the energy of the signal is:

$$u(T) = \int_{-\infty}^{+\infty} s^2(\tau)d\tau = \int_{0}^{T} s^2(\tau)d\tau = E$$

Evidently, we can also write:

$$\beta(t) = n(t)*h(t)$$

$$\beta(t) = \int_{-\infty}^{+\infty} n(\tau)s(T-t+\tau)d\tau$$

And in particular:

$$\beta(T) = \int_{-\infty}^{+\infty} n(\tau)s(\tau)\,d\tau = \int_{0}^{T} n(\tau)d\tau$$

Through the same method:

$$\beta\big((k+1)T\big) = \int_{kT}^{(k+1)T} n(\tau)d\tau$$

We can note that the noise $\beta(t)$ transformed by filtering of a Gaussian noise is still a Gaussian noise.

If $N_0/2$ is the bilateral density power of the input noise $n(t)$ and the gain of the adapted filter $G_a(f)$, then the noise power σ^2 of $\beta(t)$ is:

$$\sigma^2 = \frac{N_0}{2} \int_{-\infty}^{+\infty} |G_a(f)|^2 \, df$$

But the impulse response of $G_a(f)$ is $s(T-t)$, then:

$$G_a(f) = \overline{S}(f)e^{-2i\pi ft}$$

where $S(f)$ is the Fourier transform of $s(t)$.

Then, using the Parseval theorem:

$$\sigma^2 = \frac{N_0}{2} \int_{-\infty}^{+\infty} |S(f)|^2 \, df = \frac{N_0}{2} \int_{-\infty}^{+\infty} |s(\tau)|^2 \, d\tau = \frac{N_0 E}{2}$$

Now by considering the output of the adapted filter at the sampling time $(k+1)T$, we have:

$$y\big((k+1)T\big) = y_{k+1} = a_k E + \beta\big((k+1)T\big)$$

where we recall that $a_k = \pm 1$. Now we compare y_{k+1} to the limit S and the decision is made in such a way that:

$$\begin{cases} y_{k+1} \geq S & \text{we decide that } a_k \text{ was } +1\,(\hat{a}_k = +1) \\ y_{k+1} \leq S & \text{we decide that } a_k \text{ was } -1\,(\hat{a}_k = -1) \end{cases}$$

The representation of the decision is shown in Figure 4.6.

From the conditions of the last results, we have a more general result:

$$\begin{cases} a_k = +1 & \text{if } y_{k+1} = E + \beta \geq S \quad \Rightarrow \quad \beta \geq S - E \\ a_k = -1 & \text{if } y_{k+1} = -E + \beta \geq S \quad \Rightarrow \quad \beta \geq S + E \end{cases}$$

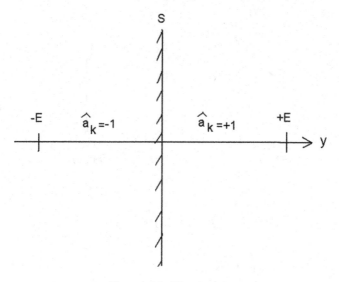

Figure 4.6. *The decision*

The total error probability is:

$$\boxed{\mathcal{P}_e = p_1 \mathcal{P}_\beta \left\{ y_{k+1} \leq S \mid a_k = +1 \right\} + p_{-1} \mathcal{P}_\beta \left\{ y_{k+1} \geq S \mid a_k = -1 \right\}}$$

where $\mathcal{P}_\beta \left\{ y_{k+1} \leq S \mid a_k = +1 \right\}$ is the probability such that $a_k = +1$ if $y_{k+1} \leq S$, the same definition occurs for the second quantity; p_1 and p_{-1} are the weight functions and are characteristic of the source and the receptor; S is the threshold of the decision and $\mathcal{P}_\beta$ is the distribution of probability of the noise β.

The total error $\mathcal{P}_e$ can be written as:

$$\boxed{\mathcal{P}_e = p_1 \mathcal{P}_\beta \left\{ \beta \leq S - E \mid a_k = +1 \right\} + p_{-1} \mathcal{P}_\beta \left\{ \beta \geq S + E \mid a_k = -1 \right\}}$$

The probability distribution of the thermal noise is a Gaussian curve of variance σ (Figure 4.7).

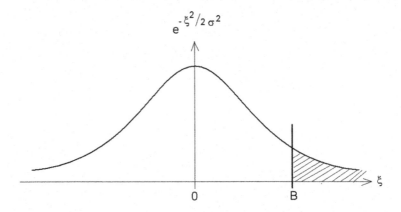

Figure 4.7. *Gaussian curve of variance σ*

$$\mathcal{P}_\beta \left[\beta \geq B \right] = \frac{1}{\sqrt{2\pi}\sigma} \int\limits_B^{+\infty} e^{-\xi^2/2\sigma^2} d\xi$$

We use to write this quantity as:

$$\mathcal{P}_\beta \left[\beta \geq B \right] = \frac{1}{2}\left(1 - erf \frac{B}{\sigma\sqrt{2}} \right)$$

With the error function:

$$erf\, X = \frac{2}{\sqrt{\pi}} \int\limits_X^{\infty} e^{-t^2} dt$$

Then we have in Figure 4.8:

$$\mathcal{P}_e = \frac{p_1}{2}\left(1 - erf \frac{E-S}{\sigma\sqrt{2}} \right) + \frac{p_{-1}}{2}\left(1 - erf \frac{E+S}{\sigma\sqrt{2}} \right)$$

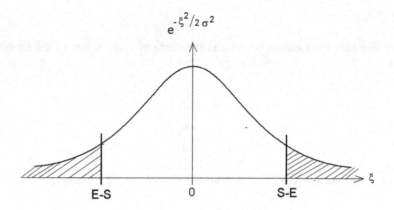

Figure 4.8. *The error function*

This error probability is minimum if:

$$\frac{d\mathscr{P}_e}{dS} = 0 \text{ or } S = \frac{\sigma^2}{2E}\log\frac{p_{-1}}{p_1}$$

If the two states have the same probability $p_1 = p_{-1} = \frac{1}{2}$, the threshold is zero $S = 0$ and:

$$\boxed{\mathscr{P}_e = \frac{1}{2}\left(1 - erf\frac{E}{\sigma\sqrt{2}}\right)}$$

And using $\sigma^2 = N_0 E/2$, we have the important result:

$$\boxed{\mathscr{P}_e = \frac{1}{2}\left(1 - erf\sqrt{\frac{E}{N_0}}\right)}$$

which gives the error below which is incorrect.

Or which is a more physical result for the telecommunication specialists:

$$\boxed{\mathscr{P}_e = \frac{1}{2}\left(1 - erf\sqrt{\frac{P_s}{P_n}}\right)}$$

4.2.2. S/N ratio

We consider the medium power and the medium noise power of the numerical signal:

$$\begin{cases} P_s = \dfrac{E}{T} \\ P_n = \dfrac{N_0}{T} \end{cases}$$

The error probability is given in function of P_s/P_n or the S/N ratio:

$$\frac{S}{N}(\text{dB}) = 10\log\frac{P_s}{P_n}$$

If $S/N = -\infty$, then $P_s = 0$ and $\mathscr{P}_e = \frac{1}{2}$. The possibilities to have $a_k = +1$ or $a_k = -1$ are the same, which means that the receptor receives only noise and no information.

On the contrary, when $S/N = +\infty$, then $P_n = 0$ and $\mathscr{P}_e = 0$. There are no errors.

In the telecommunications domain, a good S/N ratio is approximately 7–8 dB (which corresponds to an error probability of $\mathscr{P}_e \approx 10^{-3} - 10^{-4}$). Below 7 dB the quality of the transmission decreases.

	S/N dB	$\mathscr{P}_e$
bad	5	$1.2 \; 10^{-2}$
	6	$2.4 \; 10^{-3}$
minimum	7	$.77 \; 10^{-3}$
	8	$1.9 \; 10^{-4}$
	10	$3.8 \; 10^{-6}$
good	12	$.95 \; 10^{-9}$
	14	$.60 \; 10^{-12}$

Figure 4.9. *Some values of the S/N ratio*

We give some values of S/N ratio and of the error probability $\mathcal{P}_e$ (Figure 4.9).

We can conclude from Figure 4.10 that the R.R. receives a noisy sequence $\{a_k\}$, which is estimated with an error probability $\mathcal{P}_e$ and gives a noiseless sequence $\{\hat{a}_k\}$ at the output.

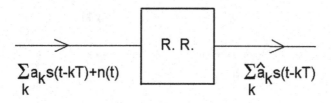

$$\sum_k a_k s(t-kT)+n(t) \qquad\qquad \sum_k \hat{a}_k s(t-kT)$$

Figure 4.10. *The work of the repeater regenerator*

The model of the adapted filter was considered as an ideal circuit. But this adapted filter is realized with difficulty. We also have to consider the distortions of the different supports (such as radio beams, satellites, optic fibers and Radars). This is very important and that is why we will discuss inter-symbol interferences (I.I.).

4.2.3. Loss support—inter-symbol interferences—eye pattern

For simplification purposes, the different supports are considered as a microwave line with losses. The channel is a dispersive environment and its properties vary with the frequency. From the classical theory of the lines (Figure 4.11), we can give the propagation constant γ and the characteristic impedance Z_c of a line of dx length:

$$\gamma = \sqrt{(R+jL\omega)(G+jC\omega)}$$

$$Z_c = \sqrt{\frac{R+jL\omega}{G+jC\omega}}$$

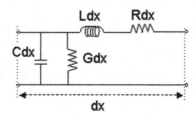

Figure 4.11. *Loss line*

In the case of weak losses:

$$\begin{cases} R \lessless L\omega \quad \text{and} \quad G \lessless C\omega \\ \gamma = \dfrac{1}{2}\left(GZ_c + \dfrac{R}{Z_c}\right) + j\omega\sqrt{LC} = \alpha + j\beta \end{cases}$$

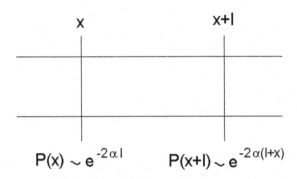

Figure 4.12. *Power of the loss line*

After a distance of l, the voltages and the currents are multiplied by the factor:

$$e^{-j\beta l - \alpha l}$$

But the powers (Figure 4.12) are multiplied by $e^{-2\alpha l}$. Then, the attenuation coefficient is a function of the power in the x plane and that in the $(x + l)$ plane.

$$\alpha = \frac{1}{2l} \log \frac{P(x)}{P(x+l)}$$

Then we have the method to determine the attenuation of the different supports.

This gives two kinds of attenuation:

– the first part that can be assimilated to "the skin effect" in the conductors, where the resistance R varies as $\sqrt{f}$;

– the second part that can be assimilated to "the leakage effect" in the dielectric, where the conductance G varies as f.

$$\alpha = k\sqrt{f} + k'f$$

and:

$$H(f) = e^{-\alpha l - j\beta l}$$

give:

$$\boxed{H(f) = e^{-(K\sqrt{f} + K'f) - j2\pi\theta f}}$$

where $K = kl$, $K' = k'l$, $\beta l = \dfrac{\omega}{v} l = \dfrac{2\pi l}{v} f = 2\pi\theta f$.

The first term is the amplitude that depends on the frequency f whereas the second term is the phase of the ideal support.

The response of the support $r(t)$ to an impulsion $x(t) = A\,Square_{T/2}(t)$ of duration T and of amplitude A is shown in Figure 4.13.

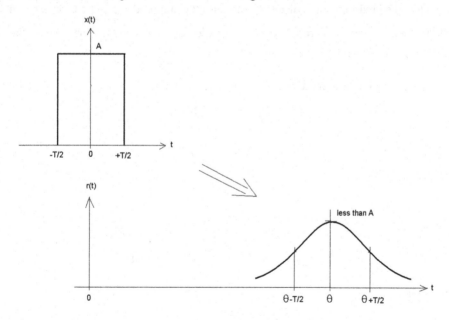

Figure 4.13. *Response of the support to a pulse*

The recovered signal $r(t)$ is:

$$\begin{cases} r(t) = x(t) * h(t) = TF^{-1}\left[X(f) \cdot H(f)\right] \\ \text{with } X(f) = AT\dfrac{\sin \pi fT}{\pi fT} \end{cases}$$

Then:

$$\boxed{r(t) = \int_{-\infty}^{+\infty} X(f) \cdot H(f) e^{j2\pi f t} df}$$

with:

$$H(f) = e^{-\left(K\sqrt{f} + K'f\right) - j2\pi\theta f}$$

Then the environment is dispersive because the amplitude of $H(f)$ depends on the frequency f. And there will be a deformation of the result $r(t)$, the smaller the T, the greater the deformation.

We can remark that if the environment is not dispersive:

$$H(f) = e^{-\alpha - j2\pi\theta f}$$

and:

$$h(t) = e^{-\alpha}\delta(t - \theta)$$

then:

$$r(t) = x(t) * h(t) = e^{-\alpha}\left[x(t) * \delta(t - \theta)\right]$$

which gives:

$$\boxed{r(t) = e^{-\alpha}x(t - \theta)}$$

which is a perfect square translated of the quantity θ and of amplitude $Ae^{-\alpha}$ (Figure 4.14).

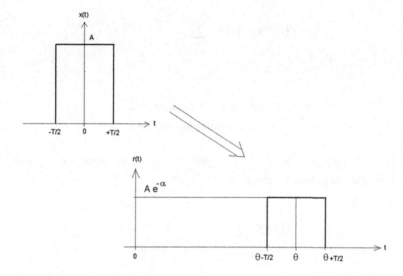

Figure 4.14. *Non-dispersive channel*

Now suppose we have to transmit a pulse train. At the output, the other symbols and particularly the neighboring symbols will have an influence on one transmit signal (Figure 4.15). This is the problem of the I.I.

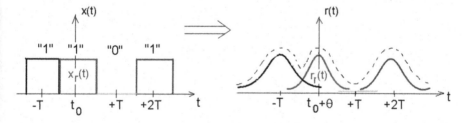

Figure 4.15. *Transmission of a pulse train on a noisy channel. For a color version of the figure, see www.iste.co.uk/jarry/communications.zip*

For example, the pulse $x_r(t)$ after transmission will be $r_r(t)$ and extends outside the interval $[t_0 + \theta - T/2, t_0 + \theta + T/2]$.

And if we have a pulse train, we must consider the whole response that takes into account the influences of all the pulses:

$$r(t) = c_0 r_r(t) + \sum_{\substack{i \neq 0,\, i=-\infty}}^{+\infty} c_i r_r(t - iT)$$

where:

$$c_i = \begin{cases} 1 & \text{if the pulse number } i \text{ is } 1 \\ 0 & \text{if the pulse number } i \text{ is } 0 \end{cases}$$

In practice, only a finite number of pulses have an influence on the value of $r(t)$. The adjacent pulses are given when $i = 1$:

$$\boxed{r(t) \approx c_0 r_r(t) + \sum_{\substack{i \neq 0,\, i=-m}}^{+m} c_i r_r(t - iT)}$$

The result $r(t)$ at the time $t = t_0 + \theta$ is a function of all the other intervals and then of the others $c_i, i \neq 0$.

If $c_0 = 1$, the emitting pulse was "1" and there exists a combination of $c_i = 1$ or 0 with $i \neq 0$ that minimizes $r(t)$. As shown in Figure 4.16, we call this suite $r_+(t)$.

$$r_+(t) \approx Inf \left\{ c_0 r_r(t) + \sum_{\substack{i \neq 0,\, i=-m}}^{+m} c_i r_r(t - iT) \right\}$$

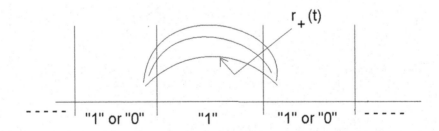

Figure 4.16. *Minimization of* $r(t)$

Using the same method, if $c_0 = 0$ then the emitting pulse was "0" and there exists a combination of $c_i = 1$ or 0 with $i \neq 0$ that maximizes $r(t)$. As shown in Figure 4.17, we call this suite $r_-(t)$.

$$r_-(t) \approx Sup\left\{ c_0 r_r(t) + \sum_{i \neq 0, i=-m}^{+m} c_i r_r(t - iT) \right\}$$

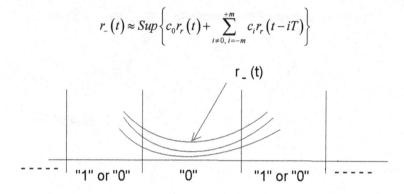

Figure 4.17. *Maximization of* $r(t)$

The association of the two curves $r_+(t)$ and $r_-(t)$ gives the eye pattern (Figure 4.18), which is a periodic function of T.

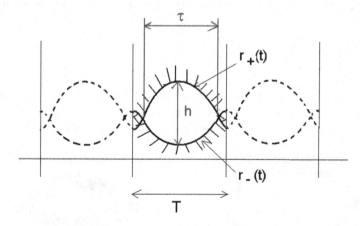

Figure 4.18. *The eye pattern. For a color version of the figure, see www.iste.co.uk/jarry/communications.zip*

$h(t) = r_+(t) - r_-(t)$ and τ are, respectively, the height and the opening of the eye. These quantities are typical characteristics of the considered transmission system and they are easily observed using a scope synchronized by the frequency clock (Figure 4.19).

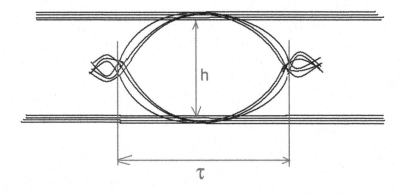

Figure 4.19. *Eye pattern on the scope. For a color version of the figure, see www.iste.co.uk/jarry/communications.zip*

If the I.I. is very strong, the filtering of the support is very significant and the eye is completely closed. Even without noise there exist configurations that are not recognized without an important error. These systems of transmission must be eliminated.

4.2.4. Noisy support

The additional noise will bring confusion near the edge of the eye, and this confusion is further extended in the eye center (Figure 4.20). The scope observation allows us to appreciate the "eye opening" and then the quality of the transmission.

open eye   closed eye

Figure 4.20. *Noisy eye patterns on the scope*

4.3. Equalization of the transfer function of the support

The influence of the support on the pulses allows us to reduce the distortion. We put an equalizer at the input of the R.R. This preamplifier or active filter will compensate the response of the support from 0 to a cutoff frequency f_E (Figure 4.21).

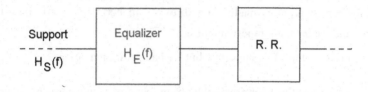

Figure 4.21. *The equalizer. For a color version of the figure, see www.iste.co.uk/jarry/communications.zip*

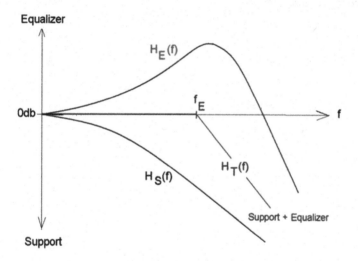

Figure 4.22. *Total gain. For a color version of the figure, see www.iste.co.uk/jarry/communications.zip*

The total transfer function is:

$$H_T(f) = H_S(f).H_E(f)$$

where we know the transfer function of the support:

$$H_S(f) = e^{-\alpha l - j2\pi\frac{l}{v}f}$$

and:

$$\left| H_S\left(f\right)\right| = e^{-\alpha l} \quad \text{with} \quad \alpha = k\sqrt{f} + k'f$$

The optimum value of f_E is determined by experiment:

– if the equalization is made from 0 to ∞ in hertz $\left(f_E = \infty\right)$, then the I.I. disappears but we have an important noise;

– if f_E is weak, the I.I. is important but we have a limited noise.

Figure 4.23 gives the error probability $\mathscr{P}_e$ in function of the cutoff frequency f_E.

Then we have to do a compromise and the optimal f_E is taken about half of the rate frequency (Figure 4.23).

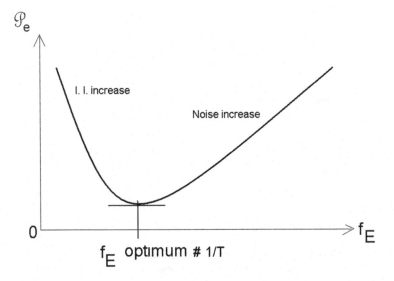

Figure 4.23. *Experimental determination of f_E*

Nyquist Criteria

5.1. Transmission without inter-symbol interference

Suppose we have to transmit a binary element "1" through the support of transfer function $H_S(f) = e^{-\alpha - j2\pi\frac{l}{v}f}$ and that this support is dispersive $\alpha = k\sqrt{f} + k'f$. The resulting pulse is shown in Figure 5.1.

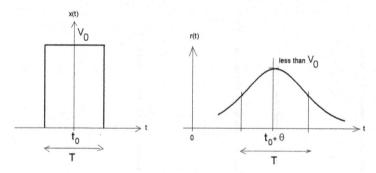

Figure 5.1. *Transmitting through the support* $H_S(f)$

According to the Nyquist criteria, "the absence of I.I. does not involve the resultant signal $r(t)$ that is of duration less than T".

In fact, it is sufficient that this resultant signal $r(t)$ will be:

– different from zero at the sampling frequency $r(t_0 + \theta) \neq 0$;

– zero at the other sampling frequency time $r(t_0 + \theta + kT) = 0$.

Then the adjacent pulses have no influence on the sampling time of the considered pulse. This result will be shown by three Nyquist criteria.

5.2. First Nyquist criterion

5.2.1. *Theoretical input message composed of Dirac pulses*

Let us consider a numerical message composed of Dirac pulses:

$$x_0(t) = \sum_{-\infty}^{+\infty} a_k \delta(t - kT)$$

In particular, we consider the NRZ signal with $a_0 = a_2 = a_3 = a_5 = 1$ and $a_1 = a_4 = 0$ (Figure 5.2).

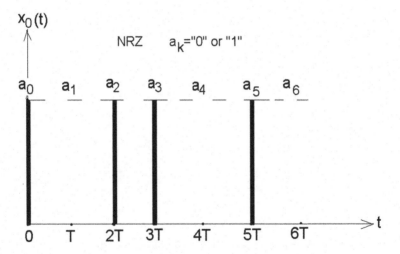

Figure 5.2. *A signal composed of Dirac pulses*

Now let us consider the rectangular filter of Figure 5.3.

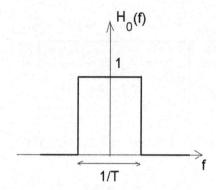

Figure 5.3. *Rectangular filter of unit amplitude*

The pulse response of this filter is:

$$h_0(t) = \int_{-1/2T}^{+1/2T} e^{2j\pi ft} df = \frac{1}{T} \frac{\sin(\pi/T)/t}{\pi t/T}$$

or:

$$\boxed{h_0(t) = \frac{1}{T} \mathrm{Sin}\, c \frac{\pi t}{T}}$$

The response $y_0(t)$ of $x_0(t)$ through $h_0(t)$ is shown in Figure 5.4.

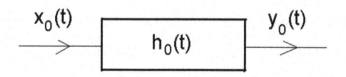

Figure 5.4. $x_0(t)$ *through* $h_0(t)$

and:

$$y_0(t) = x_0(t) * h_0(t) = \sum_{-\infty}^{+\infty} a_k h_0(t - kT)$$

This gives:

$$y_0(t) = \sum_{-\infty}^{+\infty} \frac{a_k}{T} \frac{\sin \dfrac{\pi(t-kT)}{T}}{\dfrac{\pi(t-kT)}{T}}$$

and if we sample $y_0(t)$ at the moments: $0, T, 2T, 3T, \ldots$ (Figure 5.5), we note that:

$$y_0(kT) = \frac{a_k}{T}$$ where $a_k = 0$ or 1

and we have no I.I.

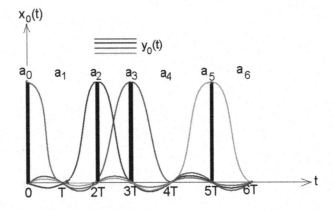

Figure 5.5. *Samples of the signal. For a color version of the figure, see www.iste.co.uk/jarry/communications.zip*

We can recover the input signal by multiplying with a Dirac function (Figure 5.6):

$$r_0(t) = y_0(t) \cdot \sum_{-\infty}^{+\infty} \delta(t-kT)$$

or after putting the function $y_0(t)$ into the sum Σ :

$$r_0\left(t\right)=\sum_{-\infty}^{+\infty}y_0\left(kT\right)\cdot\delta\left(t-kT\right)=\sum_{-\infty}^{+\infty}\frac{a_k}{T}\cdot\delta\left(t-kT\right)=\frac{1}{T}x_0\left(t\right)$$

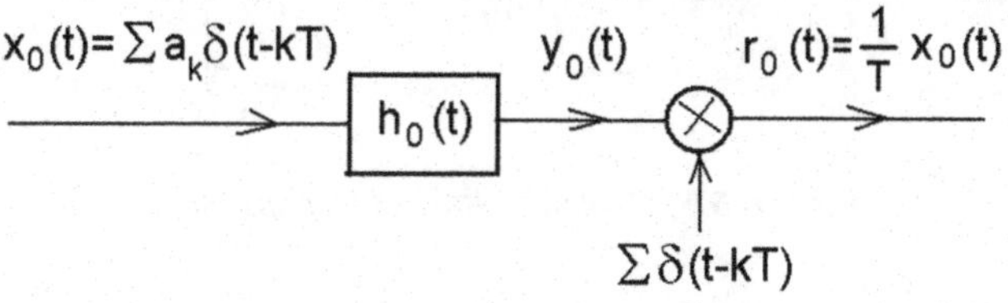

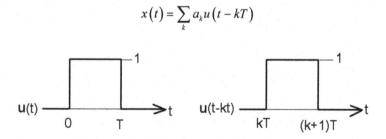

$x_0(t)=\sum a_k\delta(t-kT)$　　　$y_0(t)$　　$r_0(t)=\dfrac{1}{T}x_0(t)$

$h_0(t)$

$\sum\delta(t-kT)$

Figure 5.6. *Synoptic*

Then the absence of I.I. implies that the signal $r_0\left(t\right)$ that arrives at the sampler satisfies:

$$r_0\left(t_0+\theta\right)=\frac{1}{T}x_0\left(t+\theta\right)\neq0$$

and is zero at other sampling times.

5.2.2. Realistic message composed of elementary pulse of T duration

Now let us consider a message composed of an elementary square pulse of T duration (Figure 5.7):

$$x\left(t\right)=\sum_{k}a_k u\left(t-kT\right)$$

u(t)　　　　　1　　　　　　　u(t-kt)　　　　　1

0　　　T　　　　　　　　　　　　　kT　　　(k+1)T

Figure 5.7. *Elementary pulse of T duration*

We deal with this problem by using a realistic chain of transmission (Figure 5.8).

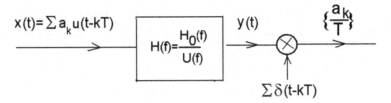

Figure 5.8. *Realistic chain of transmission*

Now $x(t)$ goes through a filter with a transfer function:

$$H(f) = \frac{H_0(f)}{U(f)}$$

If $u(f)$, $x(f)$ and $y(f)$ are the Fourier transform of $u(t)$, $x(t)$ and $y(t)$ we have:

$$X(f) = \sum_k a_k e^{-j2\pi fkt} U(f) = U(f) \cdot \sum_k a_k e^{-j2\pi fkt}$$

and:

$$Y(f) = H(f) \cdot X(f) = H_0(f) \cdot \sum_k a_k e^{-j2\pi fkt}$$

This means:

$$Y(f) = \sum_k a_k e^{-j2\pi fkt} \cdot H_0(f)$$

then:

$$\boxed{y(t) = \sum_k a_k h_0(t - kT) = \sum_k \frac{a_k}{T} \operatorname{Sin} c \frac{\pi(t - kT)}{T}}$$

After a sampling, we recover all the elements:

$$\left\{ \frac{a_k}{T} \right\}$$

5.2.3. Application to a dispersive support

The results are also valuable in the case of a dispersive support defined by a transfer function $H_S(f)$. We just have to add an equalizer $H_E(f)$.

The equalizer must be:

$$H_E(f)\cdot H_S(f)\cdot U(f) \equiv H_0(f)$$

The product $H_E(f)\cdot H_S(f)$ corresponds to the amplitude equalization.

Then we have to use:

$$H_E(f) = \frac{H_0(f)}{H_S(f)\cdot U(f)}$$

Then we also get:

$$y(t) = \sum_k \frac{a_k}{T} Sinc\frac{\pi(t-kT)}{T}$$

and after a sampling, we have the sampling elements:

$$\left\{\frac{a_k}{T}\right\}$$

and:

$$\{a_0, a_1, a_2, a_3, ...\}$$

5.2.4. Difficulty of the implementation

However, it is not possible to implement the first Nyquist criterion.

First, the filter $H_0(f)$ is an ideal filter. Moreover, its pulse response:

$$h_0(t) = \frac{1}{T} Sin c\frac{\pi t}{T} = \frac{1}{T}\frac{\sin\frac{\pi}{T}t}{\frac{\pi}{T}t}$$

must have an eye absolutely closed except at the sampling moments. At these perfect times, we have:

$$y_0(kT) = \frac{a_k}{T}$$

but there is a fluctuation of the clock frequency $kT \rightarrow kT + \varepsilon$ and:

$$y_0(kT + \varepsilon) = \frac{a_k}{T} \frac{\sin\dfrac{\pi\varepsilon}{T}}{\dfrac{\pi\varepsilon}{T}} + \sum_{l \neq 0} \frac{a_{k+l}}{T} \frac{\sin\dfrac{\pi(l+\varepsilon)}{T}}{\dfrac{\pi(l+\varepsilon)}{T}}$$

The second term is not bound, the series is then divergent and we have I.I.

The first Nyquist criterion does not tolerate any small error on the sampling frequency at the reception. This is why we give the second Nyquist criterion.

5.3. Second Nyquist criterion

The absence of I.I. is still verified if we consider a transfer function of the form:

$$H_1(f) = \frac{1}{2}(1 + g(f))$$

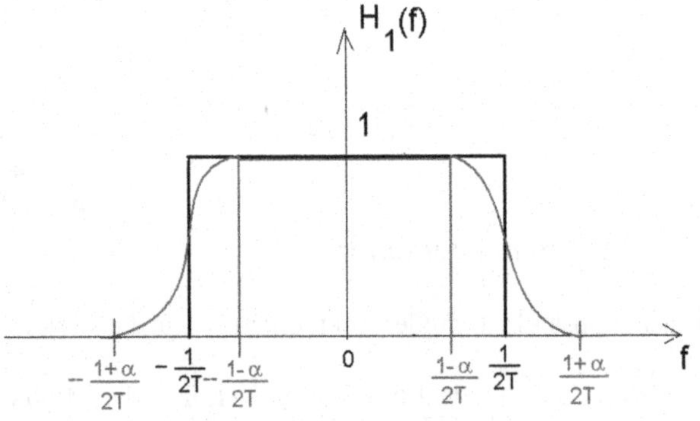

Figure 5.9. *Realist filter. For a color version of the figure, see www.iste.co.uk/jarry/communications.zip*

There is an infinite number of solutions and, in general, we use the filter as shown in Figure 5.10, which is called a "raised cosine".

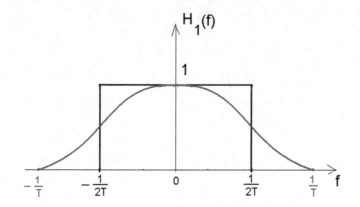

Figure 5.10. *Raised cosine. For a color version of the figure, see www.iste.co.uk/jarry/communications.zip*

The raised cosine is a very simple solution:

$$H_1(f) = \frac{1}{2}\left(1 + \cos\frac{\pi f}{T}\right)$$

and with an impulse response of the form:

$$h_1(t) = \frac{1}{T}\frac{\sin\frac{\pi}{T}t}{\frac{\pi t}{T}} \cdot \frac{\cos\frac{\pi t}{T}}{1-\left(\frac{2t}{T}\right)^2} = h_0(t) \cdot \frac{\cos\frac{\pi t}{T}}{1-\left(\frac{2t}{T}\right)^2}$$

This function is zero for all values of $kT/2$ except for 0 and the first $T/2$. Figure 5.11 gives a comparison between $h_1(t)$ and $h_0(t)$.

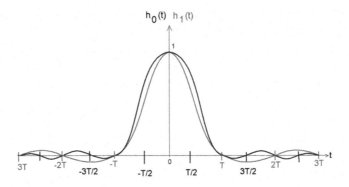

Figure 5.11. *Comparison of $h_0(t)$ and $h_1(t)$. For a color version of the figure, see www.iste.co.uk/jarry/communications.zip*

$h_1(t)$ verifies the first Nyquist criterion and gives us the elements $\{a_k/T\}$ with:

$$y_0(kT) = \frac{a_k}{T}$$

Moreover, the I.I. due to the variations of the frequency clock are reduced, which is due to the term:

$$\frac{\cos\dfrac{\pi t}{T}}{1 - \left(\dfrac{2t}{T}\right)^2}$$

which produces the series $y_0(kT + \varepsilon)$ convergent.

As it permits a tolerance on the samples, this second Nyquist criterion is more realistic than the first criterion. In addition, it is possible to show that the eye is not closed.

5.4. Third Nyquist criterion

Figure 5.12 shows a detection by using first integration, after a sample at time $t = T$ and then a reset.

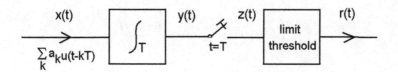

Figure 5.12. *Detection by integration*

$h(t)$ is the pulse response of the filter $H(f)$ as shown in Figure 5.13:

$$H(f) = \frac{\pi f T}{\sin \pi f T} \cdot rect\, f T$$

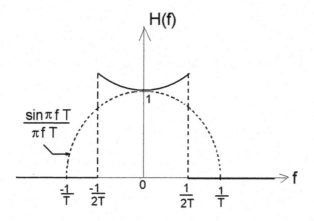

Figure 5.13. *Transfer function of* $H(f) = \frac{\pi f T}{\sin \pi f T} \cdot rect\, f T$

The integral of this pulse response is given by:

$$H(f) \cdot \frac{\sin \pi f T}{\pi f T} = rect\, fT$$

By taking the inverse Fourier transform of this equality, we have:

$$h(t) * \frac{1}{T} rect \frac{t}{T} = \frac{1}{T} \frac{\sin \pi t / T}{\pi t / T}$$

or:

$$\int_{-\infty}^{+\infty} h(\theta)\cdot rect\frac{\theta-t}{T}d\theta = \frac{\sin \pi t/T}{\pi t/T}$$

However, the transfer function of $rect\dfrac{\theta-t}{T}$ is of the form:

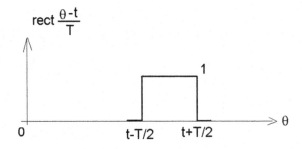

Figure 5.14. *Transfer function of* $rect(\theta-t)/T$

Then:

$$\int_{-\infty}^{+\infty} h(\theta)d\theta = \frac{\sin \pi t/T}{\pi t/T} = \begin{cases} 1 & \text{if } t=0 \\ 0 & \text{if } t=kT, k \neq 0 \end{cases}$$

More generally, the pulse response $h(t)$ satisfies the third Nyquist criterion if we have:

$$\int_{(2k-1)\frac{T}{2}}^{(2k+1)\frac{T}{2}} h(\theta)d\theta = \begin{cases} 1 & k=0 \\ 0 & k \neq 0 \end{cases}$$

and in particular:

$$\int_{-\frac{T}{2}}^{\frac{T}{2}} h(\theta)d\theta = 1$$

Then:

$$y(t) = x(t) * h(t)$$

That is:

$$y(t) = \sum_k a_k u(t - kT) * h(t) = \sum_k a_k \cdot \int_{-\infty}^{+\infty} u(t - kT - \theta) \cdot h(\theta) d\theta$$

However, $u(t - kT - \theta)$ is zero except between $kT - t$ and $kT - t + T$:

$$y(t) = \sum_k a_k \cdot \int_{kT-t+T}^{kT-t} h(\theta) d\theta$$

The integral is always zero except at the moment:

$$t = kT + \frac{T}{2}$$

and the sample is made at this moment:

$$y\left(t + \frac{T}{2}\right) = a_k \cdot \int_{+\frac{T}{2}}^{-\frac{T}{2}} h(\theta) d\theta = a_k$$

Then:

$$\boxed{z\left(t + \frac{T}{2}\right) = a_k}$$

However, the next transfer function is difficult to realize:

$$H(f) = \frac{\pi f t}{\sin \pi f t} \cdot rect\ fT$$

and we can only approach the third Nyquist criterion. Elimination of the I.I. is not perfect.

5.5. Error probability of the whole line

Let us consider the link formed by n sections (Figure 5.15).

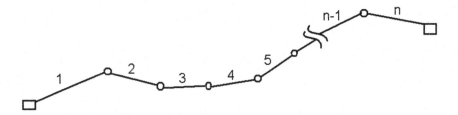

Figure 5.15. *The whole link*

$\mathscr{P}_1$ is the error probability of section $n°1$; $\mathscr{P}_2$ is the error probability of section $n°2$, ... and $\mathscr{P}_e$ is the error probability of the link formed by n sections.

Suppose that only one binary element is sent through the link. It will be false if it is recognized false on an odd number of sections.

The probabilities $\mathscr{P}_1$, ..., $\mathscr{P}_i$ are independent and the probability of having a false (j) transmission and $(n-1)$ correct transmissions is:

$$\boxed{\pi_1 = \sum_{j=1}^{n} \mathscr{P}_j \prod_{\substack{k=1 \\ k \neq j}}^{n} (1 - \mathscr{P}_k)}$$

We give the probability of having:

– $2p+1$ false transmissions;

– $n-2p-1$ correct transmissions.

$$\boxed{\pi_{2p+1} = \sum_{\substack{j_1,j_2,\dots,j_{2p+1} \\ j_1 \neq j_2 \neq --- \neq j_{2p+1}}} \mathscr{P}_{j1} \cdot \mathscr{P}_{j2} \cdots \mathscr{P}_{j2p+1} \cdot \pi_{k \neq j_p} (1 - \mathscr{P}_k)}$$

and the total error probability of the link formed by n sections is:

$$\boxed{\mathscr{P}_e = \sum_{1 \leq 2l+1 \leq n} \pi_{2l+1}}$$

As examples we consider the simple cases where $n = 2$ and $n = 3$:

$$n = 2 \quad \mathscr{P}_e = \mathscr{P}_1\left(1 - \mathscr{P}_2\right) + \mathscr{P}_2\left(1 - \mathscr{P}_1\right)$$

$$n = 3 \quad \mathscr{P}_e = \mathscr{P}_1\left(1 - \mathscr{P}_2\right)\left(1 - \mathscr{P}_3\right) + \mathscr{P}_2\left(1 - \mathscr{P}_3\right)\left(1 - \mathscr{P}_1\right) + \mathscr{P}_3\left(1 - \mathscr{P}_1\right)\left(1 - \mathscr{P}_2\right) + \mathscr{P}_1\mathscr{P}_2\mathscr{P}_3$$

We can generalize by considering a constant error probability $\mathscr{P}$ on the different sections:

$$\boxed{\mathscr{P}_e = \sum_{1 \le 2l+1 \le n} C_n^{2l+1} \mathscr{P}^{2l+1} \left(1 - \mathscr{P}\right)^{n-2l+1}}$$

If n is not as large and if the error probability $\mathscr{P}$ is limited then:

$$\boxed{\mathscr{P}_e \approx n\,\mathscr{P}}$$

or, in general, if the error probabilities are different:

$$\boxed{\mathscr{P}_e \approx \sum_{j=1}^{n} \mathscr{P}_j}$$

then an error probability of a link made up of 10 sections will be about 10^{-6} if the sections have an error probability of 10^{-7}.

However, the error probability is very bad if $\mathscr{P}_e$ is more or about 10^{-3}.

Then we can say that for an error probability of all the sections less than 10^{-5}, the quality of the transmission is "independent" of the number of these sections. This is a characteristic of the numerical systems.

Analog-to-Digital Conversion: Compression and Extension

6.1. Compression and coding

We have to transmit an analog wave that is coded. Then we introduce an error that can be considered as a noise: the quantization noise. If the quantization samples are of equal amplitude, then the smaller the signal, the more significant the relative error (Figure 6.1).

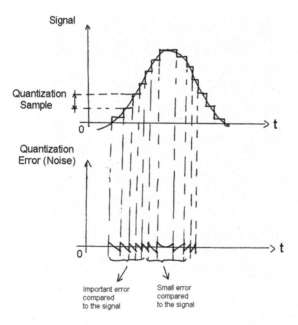

Figure 6.1. *Quantization error*

Then we need a compression of the signal. Compression is coding using several levels: we need more levels in the case of small values of the signal. The compression law will be defined later in the chapter, but here we can give the diagram of a telecommunication system (Figure 6.2).

EMISSION   **RECEPTION**

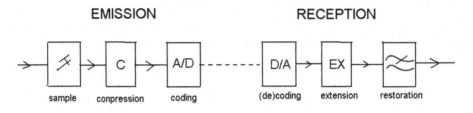

sample   conpression   coding   (de)coding   extension   restoration

Figure 6.2. *A transmission system*

6.2. Quantization

The operation of quantization is associated with the samples of a continued signal at some discrete levels; it is then an approximation.

The system of quantization is able to operate between the two values V_{MAX} and V_{MIN} (Figure 6.3).

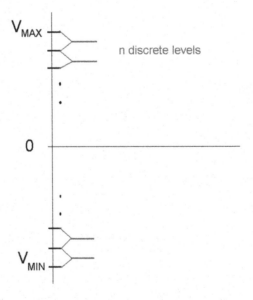

Figure 6.3. *Quantization's maximum and minimum values. For a color version of the figure, see www.iste.co.uk/jarry/communications.zip*

Using 2 bit coding (2 moments), we have the possibilities:

$$00 \quad 01 \quad 10 \quad 11$$

This gives $2^2 = 4$ discrete levels.

Coding done using 3 bits corresponds to $2^3 = 8$ possibilities of discrete levels.

In general, we use more than 3 bits but the reasoning is the same. For example, with 8 bits we have $2^8 = 256$ possibilities of discrete levels (Figure 6.4)!

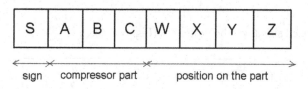

Figure 6.4. *Levels of an 8 bit coding*

The first bit S is reserved for the sign of the level; the bits A, B and C indicate the levels of the compressor and the bits W, X, Y and Z are used to give the positions on these levels.

In any case, we have loss of information because of the quantization and the resulting signal will be different from the input signal, which results in a quantization error (quantization noise).

6.2.1. Non-uniform quantization

The levels of the signal input q_i and quantizing levels x_i are not constant (Figure 6.5).

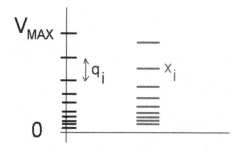

Figure 6.5. *Signal input q_i and quantizing levels x_i. For a color version of the figure, see www.iste.co.uk/jarry/communications.zip*

When $x(t)$ belongs to the interval:

$$\left[x_i - \frac{q_i}{2}, x_i + \frac{q_i}{2} \right]$$

we have the correspondence with the level x_i :

$$x(t) \rightarrow x_i$$

and the quantized signal can be written as (Figure 6.6):

$$x_Q(t) = \sum_{i=1}^{n} x_i \cdot u\left(x(t) - x_i + \frac{q_i}{2} \right) \cdot u\left(x_i + \frac{q_i}{2} - x(t) \right)$$

where $u(t)$ is the unit echelon, which is zero from $t = -\infty$ to zero and which is $+1$ from zero to $t = +\infty$.

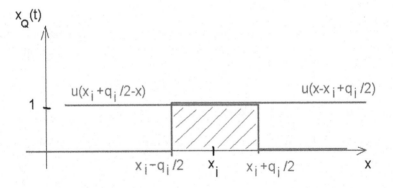

Figure 6.6. *Quantized signal* x_Q. *For a color version of the figure, see*
www.iste.co.uk/jarry/communications.zip

6.2.2. Linear quantization

The quantization is linear when:

$$q_i = q = Ct \quad \forall i$$

Then to:

$$x_i - \frac{q}{2} \leq x \leq x_i + \frac{q}{2}$$

we make the correspondence (Figure 6.7):

$$\boxed{x_i = \left(i + \frac{1}{2}\right)q}$$

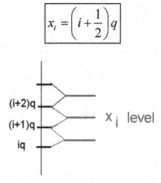

Figure 6.7. *Linear quantization. For a color version of the figure, see www.iste.co.uk/jarry/communications.zip*

This is the European law with a mid-riser form (Figure 6.8). The quantization error is (relatively) weak with a maximum of $q/2$:

$$\boxed{\left|x(t) - x_i\right| \leq \frac{q}{2}}$$

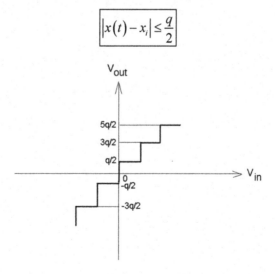

Figure 6.8. *European law*

We give in the same diagram (Figure 6.9) the restored European signal starting from the real signal. We also give the restored American signal.

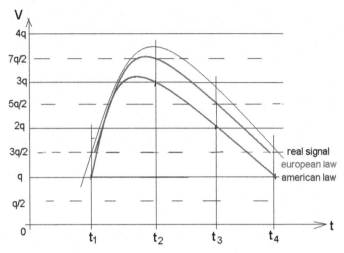

Figure 6.9. *Restored signal. For a color version of the figure, see www.iste.co.uk/jarry/communications.zip*

The American law has a mid-tread form. It makes the correspondence (Figure 6.10):

$$x_i = (i-1)q$$

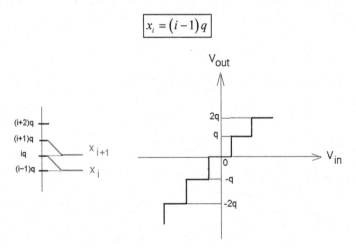

Figure 6.10. *American law. For a color version of the figure, see www.iste.co.uk/jarry/communications.zip*

The American quantization error is twice as large because:

$$\boxed{\left|x(t)-x_i\right| \le q}$$

it has a maximum equal to q and it justifies the restored signal of Figure 6.9. However, if we discuss the noise at the beginning (time off), we can see from Figure 6.11 that:

– using the European law, the noise is amplified;

– but using the American law, there is no amplification of the noise.

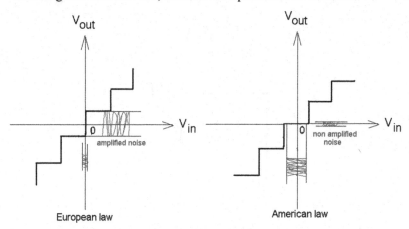

Figure 6.11. *Noise output at time off. For a color version of the figure, see www.iste.co.uk/jarry/communications.zip*

6.3. Noise's quantization

To the signal x we make the correspondence of the constant level x_i (Figure 6.12).

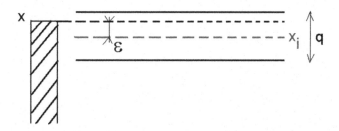

Figure 6.12. *Quantization. For a color version of the figure, see www.iste.co.uk/jarry/communications.zip*

Let us define ε as the quantization error and q as the step of quantization.

x has the same probability of taking all the values between $x_i - (q/2)$ and $x_i + (q/2)$ and the quantization error is equally distributed between $-q/2$ and $+q/2$. The density probability $p(\varepsilon)$ is then the same in the interval $[-q/2, q/2]$.

This probability is $p(\varepsilon) = 1/q$ because we must have $\int_{-q/2}^{+q/2} p(\varepsilon) d\varepsilon = 1$ (Figure 6.13).

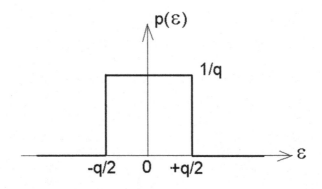

Figure 6.13. *Density probability in the interval* $[-q/2, q/2]$

The noise's power when the signal covers the distance $[-q/2, q/2]$ is:

$$\overline{E}^2 = \int_{-q/2}^{+q/2} \varepsilon^2 p(\varepsilon) d\varepsilon = \frac{1}{q} \int_{-q/2}^{+q/2} \varepsilon^2 d\varepsilon$$

This is:

$$\overline{E}^2 = \frac{1}{q} \left[\frac{\varepsilon^3}{3} \right]_{-q/2}^{+q/2} = \frac{2}{q} \frac{q^3}{3.8}$$

and the average quadratic error between the two levels is:

$$\bar{E}^2 = \frac{q^2}{12}$$

Now let us suppose that the value of the signal gives us the possibility of covering all the quantization levels (Figure 6.14).

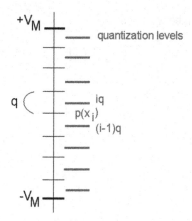

Figure 6.14. *Quantization levels. For a color version of the figure, see www.iste.co.uk/jarry/communications.zip*

In addition, suppose that $p(x_i)$ is the probability of the signal having a value between two consecutive levels $(i-1)q$ and iq. The total quadratic error is the sum of the averages:

$$\bar{E}_T^2 = \sum_i p(x_i)\frac{q^2}{12} = \frac{q^2}{12}\sum_i p(x_i)$$

The signals are not correlated:

$$p(x_i) = 1$$

and we obtain the quantization noise:

$$\bar{E}_T^2 = B_q = \frac{q^2}{12}$$

6.3.1. S/B in the case of a sinusoidal signal

In the case of a sinusoidal signal $[+V_M, -V_M]$, the power of this signal is $S = V_M/2$. In addition, if you have n bits or 2^n discreet levels and:

$$q = \frac{2V_M}{2^n}$$

the quantization noise can be written as:

$$B_q = \frac{q^2}{12} = \frac{1}{12}\left(\frac{2V_M}{2^n}\right)^2$$

and the quantity noise under power:

$$\frac{S}{B_q} = \frac{V_M^2}{2} \cdot \frac{12 \cdot 2^{2n}}{4 \cdot V_M^2}$$

and we get the ratio:

$$\boxed{\frac{S}{B_q} = \frac{3}{2} 4^n}$$

This can be written in decibels as:

$$\boxed{\left(\frac{S}{B_q}\right)_{dB} = 1.76 + 6n}$$

This shows the importance of the number of discreet levels 2^n. If we add only 1 bit, the number of discreet levels changes from 2^n to $2^{n+1} = 2 \cdot 2^n$ and is multiplied by 2, and the ratio $\left(S/B_q\right)_{dB}$ is better of 6 dB.

6.3.2. S/B in the case of a white noise

If the signal is a white noise of power (variance) σ^2, we make the same calculation by making the change:

$$S \to \sigma^2 \text{ instead of } \frac{V_M^2}{2}$$

and:

$$\frac{S}{B_q} = \frac{\sigma^2}{V_M^2} 3 \cdot 2^{2n}$$

This means:

$$\boxed{\frac{S}{B_q} = \frac{3\sigma^2}{V_M^2} \cdot 4^n}$$

The characteristics of the damage's quantization are given by an array of curves $\left(S/B_q\right)_{dB}$ in function of $\left(\sigma/V_M\right)_{dB}$, which is the power of the signal (Figure 6.15).

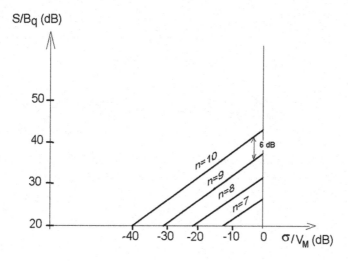

Figure 6.15. *Array of curves with different values of* n

We obtain a series of straight lines for different values of n. However, to the quantification noise, we have to add the limitation noise.

6.4. Limitation noise

Limitation noise is the charge capacity of the system. It means that we have to consider the maximum level V_M (Figure 6.16). We always restore V_M for a level x when:

$$x \geq V_M$$

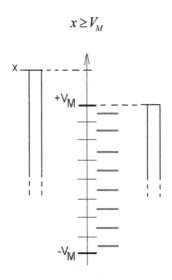

Figure 6.16. *Error of limitation. For a color version of the figure, see www.iste.co.uk/jarry/communications.zip*

There is then an error tension $x - V_M$ and the corresponding noise has a power (variance):

$$B_L = 2 \int_{V_M}^{\infty} \left(x - V_M \right)^2 . p(x) \, dx$$

where $p(x) = \dfrac{1}{\sigma\sqrt{2\pi}} e^{-x^2/2\sigma^2}$ is the density probability of a Gaussian noise and the limitation noise power is:

$$B_L = \frac{2}{\sigma\sqrt{2\pi}} \int_{V_M}^{\infty} \left(x - V_M \right)^2 e^{-x^2/2\sigma^2} \, dx$$

We make the transformation $k = V_M/\sigma$ and with $k \geq 2$ we have:

$$B_L \approx \frac{4V_M^2}{\sqrt{2\pi}} \frac{e^{-k^2/2}}{k^5}$$

and in the case of a white noise, we have a signal-to-noise ratio as:

$$\frac{S}{B_L} = \frac{\sigma^2}{B_L} = \frac{\sqrt{2\pi}}{4} k^3 e^{k^2/2}$$

with:

$$k = \frac{V_M}{\sigma}$$

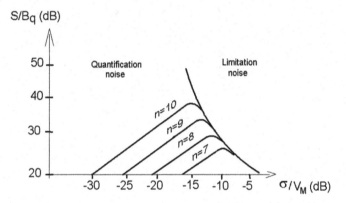

Figure 6.17. *The influence of the limitation noise*

From Figure 6.17, we can see the influence of the limitation noise:

– the S/B_q ratio increases with the increase in the bits number n;

– this ratio decreases with the decrease in the power of the input signal.

6.5. Compression and extension

To improve S/B_q ratio, we must:

– increase the bits number n;

– reduce the quantization step when the amplitude of the input signal decreases.

We cannot increase the number of bits to infinity, and we use the last possibility that is called "compression" (Figure 6.18). We can note that if we have a "compression" at the input, we must have a opposite operation at the output: an "extension".

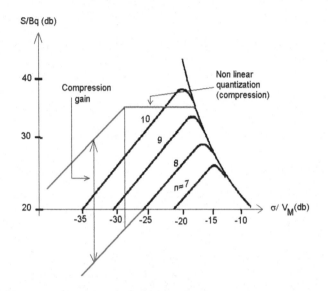

Figure 6.18. *Compression. For a color version of the figure, see www.iste.co.uk/jarry/communications.zip*

We then apply a "compression" on the emitting side, which increases with the increase in the amplitude of the input signal.

Let us determine the number of necessary bits for coding.

Without considering the limitation noise, we have:

$$\frac{S}{B_q} = \frac{\sigma^2}{V_M^2} \cdot 3 \cdot 4^n$$

Now suppose that we have a value of $S/B_q = 25$ dB, which is minimum in the case of the smallest signals (i.e. $10 \log \sigma^2/V_M^2 = -60$ dB).

Then using the value $10 \log 2 = 3$ dB, we obtain:

$$25 = -60 + 10 \log 3 + 6n$$

which means $n \approx 13$. This value is reasonable in the case of small input signals but too high in the case of high signals.

For example, in this last case consider we have only an intermediate signal so that:

$$10 \log \frac{\sigma^2}{V_M^2} = -15 \text{ dB}$$

Then:

$$\frac{S}{B_q} = -15 + 10 \log 3 + 6 \cdot 13 \approx 68 \text{ dB}$$

This value is too high. There is then the necessity of having a nonlinear quantization that is a "compression" (Figure 6.20).

Let us consider the very simple case where $n = 3$ (Figure 6.19) and of a linear quantization.

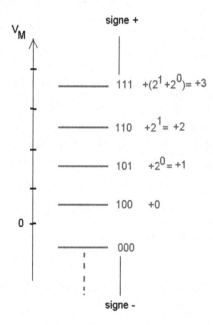

Figure 6.19. *Linear quantization for* $n = 3$. *For a color version of the figure, see www.iste.co.uk/jarry/communications.zip*

The compressor $(n = 3)$ is shown in Figure 6.20.

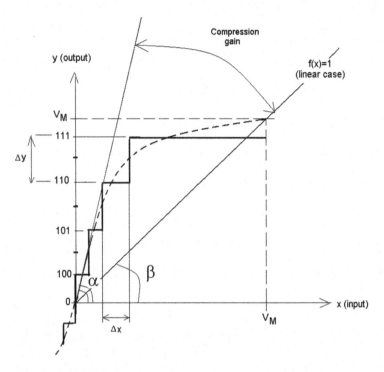

Figure 6.20. *Compression gain. For a color version of the figure, see www.iste.co.uk/jarry/communications.zip*

The output of this system $(n = 3)$ is such that Δy is a constant and Δx is given as a function of the nonlinearity $f(x)$:

$$\Delta y = \frac{2V_M}{N} = C^{nt} \quad \text{with} \quad N = 2^n$$

$$\Delta x = \frac{\Delta y}{f'(x)} = \frac{2V_M}{N} \cdot \frac{1}{f'(x)}$$

where Δx is the new step quantization that is non-constant. In the linear case, the step was a constant q.

Then we have to make the change:

$$q \rightarrow \Delta x$$

and:

$$B_q = \frac{q^2}{12} \rightarrow B_q = \frac{(\Delta x)^2}{12} = \frac{V_M^2}{3N^2} \frac{1}{(f'(x))^2}$$

Now the new S/B_q ratio is:

$$\boxed{\frac{S}{B_q} = \frac{\sigma^2}{V_M^2} = (\frac{\sigma^2}{V_M^2} \cdot 4^n \cdot 3) \cdot (f'(x))^2}$$

instead of (linear case):

$$\frac{S}{B_q} = \frac{\sigma^2}{V_M^2} = \frac{\sigma^2}{V_M^2} \cdot 4^n \cdot 3$$

If we have a logarithmic compression, the logarithmic ratio $(S/B_q)_{dB}$ is proportional to $(\sigma/V_M)_{dB}$ in decibels, and the signal-to-noise ratio is a constant.

Now we give the emitting compression law, but we have to remember that if there is an emitting compression law, there will be an extension law at the reception.

The law A is a European law and it gives the output in function of the input:

$$\begin{cases} y = \dfrac{Ax}{1 + \log_e A} & \text{when} \quad 0 \le x \le \dfrac{1}{A} \quad \text{linear part} \\ y = \dfrac{1 + \log_e Ax}{1 + \log_e A} & \text{when} \quad \dfrac{1}{A} \le x \le 1 \quad \text{nonlinear part} \end{cases}$$

where:

$$A = 87.6$$

$$x = \frac{V_{input}}{V_M} \quad \text{and} \quad y = \frac{V_{output}}{V_M}$$

The compression gain is given by the slope at the origin:

$$\frac{A}{1 + \log_e A}$$

which gives a gain of 24 dB.

The compression law has a linear part and a nonlinear part (Figure 6.21).

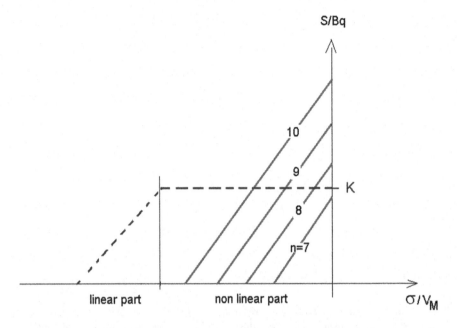

Figure 6.21. *Linear and nonlinear parts of the compression law. For a color version of the figure, see www.iste.co.uk/jarry/communications.zip*

Linear part: for technological reasons, it is impossible to reduce indefinitely the size of the steps. Then we define a minimum level at the beginning until the ratio $\left(S/B_q\right)_{dB} = K$ will take the value K.

Logarithmic part: this nonlinear part allows us to have a constant ratio $\left(S/B_q\right)_{dB} = K$ for all the values of the input $\left(\sigma/V_M\right)_{dB}$.

Then, from Figure 6.20, we define the compression rate as:

$$\tau = \frac{tg\,\alpha}{tg\,\beta}$$

In the third case $(n = 3)$, the compression law A is estimated by using $2^3 = 8$ segments of straight lines (Figure 6.22). The first two lines are collinear (this is the linear part of the law).

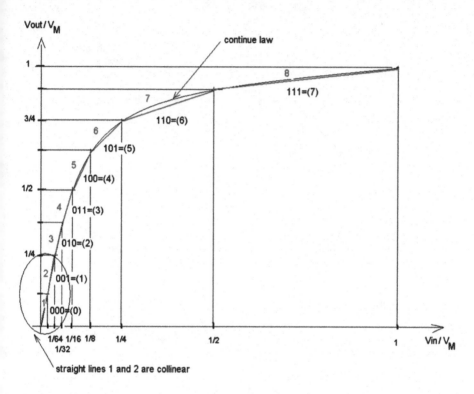

Figure 6.22. *Compression law in the case of* $n = 3$. *For a color version of the figure, see www.iste.co.uk/jarry/communications.zip*

Then law A approximated by straight lines is not a continue law. This approximation gives noise ripples that are due to the linear quantization of all the straight lines. When the ratio of the slope of two consecutive straight lines is equal to 2, then we have ripples of 6 dB (Figure 6.23). There are six ripples that correspond to the non-collinear segments: number 3 to 8.

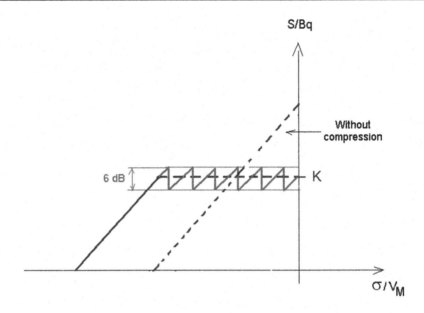

Figure 6.23. *Noise ripples. For a color version of the figure, see*
www.iste.co.uk/jarry/communications.zip

6.6. Introduction to coding

The method of coding considered is 8 bits, which allows us to have $2^8 = 256$ discrete levels (Figure 6.24).

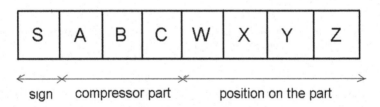

Figure 6.24. *8 bit coding*

The first bit S is reserved for the sign of the level; the bits A, B and C indicate the levels of the compressor and the bits W, X, Y and Z are used to give the positions on these levels. This gives a total of $2 \cdot 2^3 \cdot 2^4 = 2^8 = 256$ levels.

There exist several types of encoder. We will discuss only the series/parallel encoder and the "DELTA" encoder.

6.6.1. Series/parallel encoder

The encoder operations are performed in three steps:

– during the first step and using a comparator, we determine the sign of the sample. Then we have the bit S;

– during the second step, we determine the use of segments of straight line or the use of compressor. And using eight comparators, we get the bits (A, B, C);

– at the end and using 16 comparators, we get the position (W, X, Y, Z) on the compressor determined during the second step.

The encoder operations require 16 comparators because the comparators of the first and the second step use these during the third step.

This is the series/parallel encoder. Integration on a large scale is realized with microprocessors.

6.6.2. "DELTA" encoder

A DELTA (Δ) encoder compares the sample to be coded to the previous sample. This is a differential encoder (Figure 6.25).

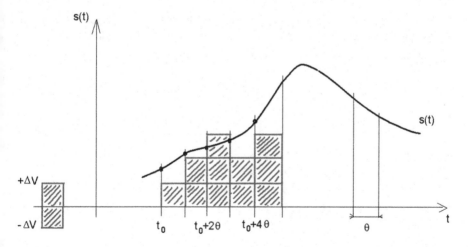

Figure 6.25. *Principle of the Δ encoder. For a color version of the figure, see www.iste.co.uk/jarry/communications.zip*

We have an analogical signal $S(t)$ and a clock, which gives a sampling frequency f_S or a time sampling $\theta = 1/f_S$. The principle is the next: if the considered sample is greater than the precedent, we emit the bit "1", on the contrary we emit the bit "0".

An example is shown in Figure 6.25.

$$
\begin{array}{lll}
\text{if} & s(t_0) \geq 0 & \text{we emit} + \Delta V \\
\text{if} & s(t_0 + \theta) \geq \Delta V & \text{we emit} + \Delta V \\
\text{if} & s(t_0 + 2\theta) \geq 2\Delta V & \text{we emit} + \Delta V \\
\text{if} & s(t_0 + 3\theta) \leq 3\Delta V & \text{we emit} - \Delta V \\
\text{if} & s(t_0 + 4\theta) \geq (3\Delta V - \Delta V) & \text{we emit} + \Delta V
\end{array}
$$

Then we get a polynomial outline (contour) that fits the signal $s(t)$.

The principle of the digitization makes the correspondence:

$$
\begin{array}{lcl}
+\Delta V & \rightarrow & "1" \\
-\Delta V & \rightarrow & "0"
\end{array}
$$

The associated numerical rate is $1/\theta = f_S$.

Moreover, the Δ modulation is possible if we dispose of the previous samples. It is possible by using a very simple RC circuit as a memory network. It furnishes at the moment t_1 or $(t_0 + \theta)$, the value of the signal at the moment t_0 (Figure 6.26).

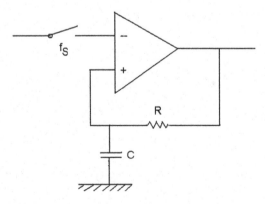

Figure 6.26. *Memory network*

This memory network is an integrator RC network and we obtain a diagram as shown in Figure 6.27.

RC output

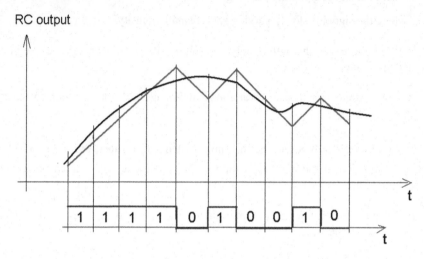

Figure 6.27. Δ *encoder with an RC memory. For a color version of the figure, see www.iste.co.uk/jarry/communications.zip*

In fact, the classical Δ modulation presents serious disadvantages because all the variations of voltage ΔV are constant.

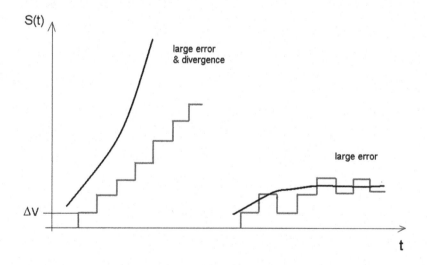

Figure 6.28. *Possibilities of encoder error. For a color version of the figure, see www.iste.co.uk/jarry/communications.zip*

This gives important errors in the two cases (Figure 6.28):

– where the signal is rapidly growing;

– where the signal is slowly varying and of small intensity.

In these two cases, the outline polygon cannot follow the real signal curve in a satisfactory manner.

An improvement is to use n variations of voltage ΔV that are not constant: $\Delta V_1, \Delta V_2, \cdots, \Delta V_n$.

The problem is now to choose the appropriated ΔV_i. In fact, this value of ΔV_i will be a function of the number of the bits "1" present during a given precedent time (10 ms for example).

A computer determines the proportion of bits "1" used in a recent time and decides what ΔV_i to use.

This is a numerical control because the value ΔV_i is controlled by the previous number of the bits "1".

Part 3

Digital Transmissions on Carriers' Frequency

Phase Shift Keying Modulations

7.1. Introduction

In order to bring more and more information, we need efficient transmission systems such as:

– radio beams (with numerical rate ranging from 50 to 100 Mbits/s);

– microwave circular wave guide (up to 35–50 GHz), which can transport more than 150,000 phone lines;

– optic fiber with laser.

In all cases, we have to transmit the binary elements "1" and "0" (sometimes also "−1") with the RZ and NRZ (or bipolar) codes.

Transmission is made by using amplitude modulation (AM), frequency modulation (FM) or phase modulation (PM).

7.2. Amplitude, frequency and phase modulation

In the case of the AM, the carrier of frequency f_0 is modulated by the bit (Figure 7.1):

$$bit\ "0"\quad \text{symbolizes}\ V_0\quad \rightarrow\quad V_0 \cos 2\pi f_0 t$$
$$bit\ "1"\quad \text{symbolizes}\ V_1\quad \rightarrow\quad V_1 \cos 2\pi f_0 t$$

This method is not used because of its high noise sensitivity. We have the same disadvantage as in the case of the AM of analog signals.

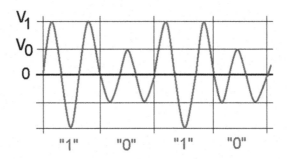

Figure 7.1. *AM modulation. For a color version of the figure, see www.iste.co.uk/jarry/communications.zip*

In the case of the FM, we have the following correspondences:

$$bit \; "0" \;\; \rightarrow \;\; carrier \; f_0$$
$$bit \; "1" \;\; \rightarrow \;\; carrier \; f_1$$

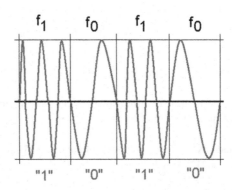

Figure 7.2. *FM modulation. For a color version of the figure, see www.iste.co.uk/jarry/communications.zip*

There is no protection against the noise as in the case of the AM.

In numerical transmission, we use the PSK. It is a PM that is well adapted to the binary aspect of the signal to be transmitted.

For all the digits "0" or "1", we associate a value to the state of the phase of the carrier of frequency f_0 during T. Then $1/T$ will be the rapidity of modulation of the system and also the numerical rate:

$$bit\ "0"\ \rightarrow\ V\cos(\omega_0 t + \varphi_0)\ during\ T\ seconds$$
$$bit\ "1"\ \rightarrow\ V\cos(\omega_0 t + \varphi_1)\ during\ T\ seconds$$

On reception, we have to distinguish φ_0 and φ_1 with the more reduced error. The maximum separation of the two phases φ_0 and φ_1 is $\varphi_1 = \varphi_0 + \pi$, and if we take the origin at $\varphi_0 = 0$, we obtain:

$$bit\ "0"\ \rightarrow\ \quad V\cos\omega_0 t\ during\ T\ seconds$$
$$bit\ "1"\ \rightarrow\ V\cos(\omega_0 t + \pi)\ during\ T\ seconds$$

The coherence imposes that the carrier frequency be a multiple of the rhythm frequency of the binary signal. If this is not the case, then the coherence is difficult to preserve and the changes of states of the phase will occur at any time in the segment $[0, 2\pi]$.

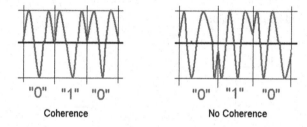

"0"   "1"   "0"   "0"   "1"   "0"

Coherence   No Coherence

Figure 7.3. *PSK modulation. For a color version of the figure, see*
www.iste.co.uk/jarry/communications.zip

In any case, instead of two values (states) of phase 0 and π, we can use a double numerical rate using four values of the phase.

For example in the case of a coherent modulation with four values of the phase, we have:

$$\varphi_0$$
$$\varphi_0 + \pi/2$$
$$\varphi_0 + \pi$$
$$\varphi_0 + 3\pi/2$$

and the numerical rate will be $2/T$ instead of $1/T$.

7.3. Phase shift keying: coherent PSK of order 2

We consider the case in which we have to transmit only two bits ("0" and "1"). The signals of the PSK are:

$$s_0(t) = a\cos(\omega_0 t + \varphi_0)$$
$$s_1(t) = a\cos(\omega_0 t + \varphi_1)$$

where $\omega_0 = 2\pi f_0$ and f_0 is the carrier frequency.

We have a coherent PSK if:

$$\begin{cases} \varphi_0 = 0 \\ \varphi_1 = \pi \end{cases}$$

The practical realization (Figure 7.4) needs an oscillator (ω_0) and four diodes (D) that are organized in a ring shape.

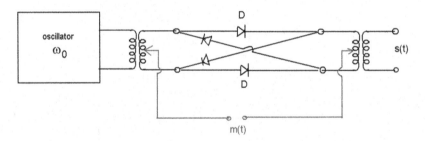

Figure 7.4. *Realization of a PSK of order 2. For a color version of the figure, see www.iste.co.uk/jarry/communications.zip*

The numerical modulated signal $m(t)$ has been symmetrized:

$$m(t) = \sum_n p_n g(t - nT)$$

with:

$$p_n = \pm 1$$

Then this practical realization can be seen as shown in Figure 7.5.

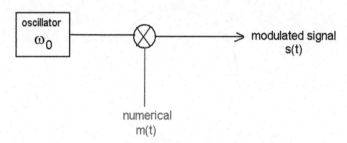

Figure 7.5. *PSK signal of order 2. For a color version of the figure, see www.iste.co.uk/jarry/communications.zip*

The result is of the form:

$$s(t) = a\cos\omega_0 t \cdot m(t) = a\sum_n p_n g(t - nT) \cdot \cos\omega_0 t$$

The numerical signal to be modulated is slightly filtered (Figure 7.6(a)) and the modulated signal is shown in Figure 7.6(b).

a) signal to be modulated

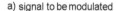
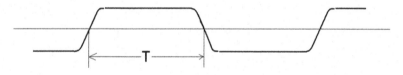

b) modulated signal PSK (coherence $\varphi_0 = 0$)

Figure 7.6. *Modulated PSK signal. For a color version of the figure, see www.iste.co.uk/jarry/communications.zip*

If T is the length of the bit, then the rapidity of the modulation is $1/T$ and the numerical rate is also $1/T$.

We also give the phases' plane of this PSK of order 2.

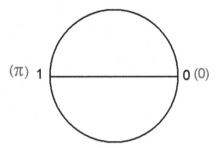

Figure 7.7. *Phases' plane of the PSK of order 2. For a color version of the figure, see www.iste.co.uk/jarry/communications.zip*

7.4. PSK of order 4

The PSK of order 4 will be realized using two squaring carriers. The elements of the numerical signal form groups of two. Then we have four possible states, and we have to use a doublet of a bit and four values of the phase of the carrier.

Suppose we have to transmit the next signal. As shown in Figure 7.8, during the first half of duration T we consider the signal A and during the second half we consider the signal B.

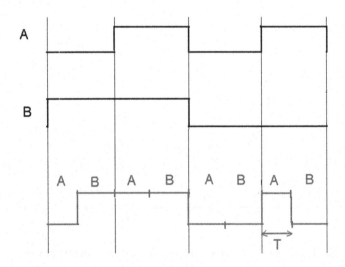

Figure 7.8. *Transmission PSK of the trains A and B. For a color version of the figure, see www.iste.co.uk/jarry/communications.zip*

We give the correspondence between the emit phase and the numerical states:

$$\left\{ \begin{array}{cccc} 00 & 01 & 11 & 10 \\ 0 & \dfrac{\pi}{2} & \pi & \dfrac{3\pi}{2} \end{array} \right.$$

We give these results in the phases' plane (Figure 7.9).

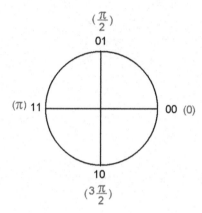

Figure 7.9. *Phases' plane of the PSK of order 4. For a color version of the figure, see www.iste.co.uk/jarry/communications.zip*

For the realization, we must consider two modulators of order 2 driven by two carriers with a phase difference of $\pi/2$. Only one oscillator is then necessary (Figure 7.10).

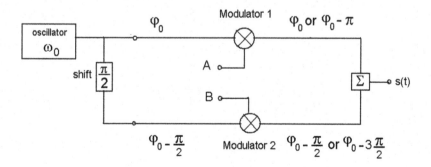

Figure 7.10. *PSK signal of order 4*

We have for the two modulators the result shown in Figure 7.11.

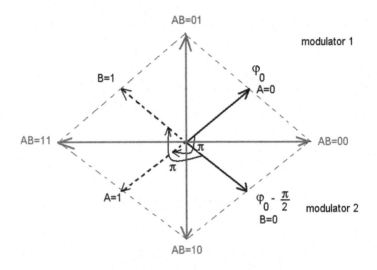

Figure 7.11. *Influence of the two modulators. For a color version of the figure, see www.iste.co.uk/jarry/communications.zip*

We used the GRAY code (Figure 7.12).

A	B	φ
0	0	0
0	1	$\dfrac{\pi}{2}$
1	1	π
1	0	$3\dfrac{\pi}{2}$

Figure 7.12. *GRAY code*

Two adjacent states are different for only one bit. When the phase shifts from π, the two bits also change.

$$00 \xrightarrow{\pi} 11$$

$$01 \xrightarrow{\pi} 10$$

Then we have the properties:

$$\varphi\left(\overline{A}, \overline{B}\right) = \varphi\left(A, B\right) + \pi$$

$$\varphi\left(A, B\right) = \varphi\left(\overline{A}, B\right) \pm \frac{\pi}{2}$$

If T is the length of the bit, then the rapidity of the modulation is $1/T$ and the numerical rate is $2/T$.

7.5. PSK of order height and 16

In the same way, we can group together the binary elements by 3 or 4 in order to obtain a PSK of order height and 16.

At each value of the triplet or the quadruplet, we make a correspondence with a value of the phase's carrier. Then we obtain a PSK modulation of order height or 16.

In addition, we have:

$$\boxed{\text{Rapidity of modulation} = \frac{\text{Numerical rate}}{3 \left(\text{or } 4\right)}}$$

There is an increase in the quantity of information, but if the number of the phase states is growing, the identification of the signal becomes more and more difficult, and the quality of the transmission decreases.

8

Differential Coding

8.1. Introduction

Coherent modulation requires:

– the exact knowledge of the phase at reception;

– the recovery of the carrier at reception.

These operations are delicate, especially at reception.

However, the differential coding determines the jumps of the carrier's phase. And we need to use a particular code to obtain the bonds of the phase, which is called transcoding.

8.2. Differential coding of order 2

The states "0" and "1" correspond to the bounds of the phase 0 and π, which means that:

$$\text{if } A = 0 \quad \text{the carrier phase is unchanged}$$
$$\text{if } A = 1 \quad \text{the carrier phase is changed from } \pi$$

as shown in Figure 8.1.

A	$\Delta\varphi$
0	0
1	π

Figure 8.1. *Evolution of the phase*

Figure 8.2 defines the logic function that makes it possible to go from signal A to command α of the modulator.

Transmitted signal	Command signal	Carrier phase
A	α	φ
0	α_0	unchanged
1	$\overline{\alpha_0}$	changed by π

Figure 8.2. *Command of the modulator*

This can be realized by using a flip-flop scale, with the clock close to numerical signal A (Figure 8.3).

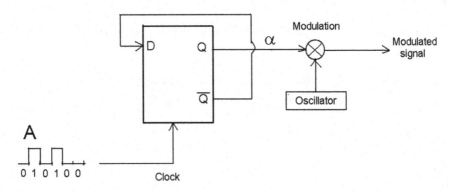

Figure 8.3. *Realization of the command of the modulator*

If we have $A = 0$, there is no clock impulsion and the flip-flop state does not change and then $\alpha = \alpha_0$ and the phase is unchanged.

However, if we have $A = 1$, the clock pulse changes the flip-flop state and $\alpha = \overline{\alpha}_0$, then the modulator inverts the carrier's phase.

8.3. Differential modulation of order 4

Starting from the four states of the doublet of information AB, there will be four corresponding bounds of the phase $\Delta\varphi$.

The phase is dependent on:

– the initial phase φ_0, which can take four values;

– the bound of the phase $\Delta\varphi$, which can also have four values.

Then we make the realization given in Figure 8.4.

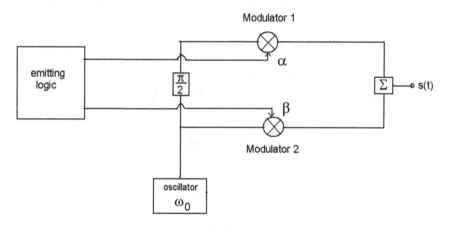

Figure 8.4. *Realization of the differential of order 4*

We have to make the differential phases $\Delta\varphi$ starting from the initial phases φ_0 shown in Figure 8.5.

A	B	Δφ
0	0	0
0	1	π/2
1	1	π
1	0	3π/2

α_0	β_0	φ_0 (initial)
0	0	0
0	1	π/2
1	1	π
1	0	3π/2

Figure 8.5. *The differential phases* $\Delta\varphi$ *and the initial phases* φ_0

As shown in Figure 8.4, the emitting logic gives the numerical signals α and β, which command the two modulators, and this forms the doublet AB of the precedent phase.

We must find the diagram that gives the bound or the doublet AB necessary to consider the previous carrier's phase $\alpha_0\beta_0$. The mechanism of the modulators is shown in Figure 8.6.

φ_0 \ $\Delta\varphi$ (AB) ($\alpha_0\beta_0$)	0 AB=00	π/2 AB=01	π AB=11	3π/2 AB=10
0 $\alpha_0\beta_0$=00	0 αβ=00	π/2 αβ=01	π αβ=11	3π/2 αβ=10
π/2 $\alpha_0\beta_0$=01	π/2	π	3π/2	0
π $\alpha_0\beta_0$=11	π	3π/2	0	π/2
3π/2 $\alpha_0\beta_0$=10	3π/2	0	π/2	π

Figure 8.6. *Mechanism of the modulators. For a color version of the figure, see www.iste.co.uk/jarry/communications.zip*

From this diagram, we can see that we have effectively on the phase:

$$\varphi(\alpha,\beta) = \varphi_0(\alpha_0,\beta_0) + \Delta\varphi(A,B)$$

In addition, we get the logical domain:

$$\alpha = \alpha_0\overline{A}\overline{B} + \beta_0\overline{A}B + \overline{\alpha}_0AB + \overline{\beta}_0A\overline{B}$$
$$\beta = \beta_0\overline{A}\overline{B} + \overline{\alpha}_0\overline{A}B + \overline{\beta}_0AB + \alpha_0A\overline{B}$$

and these expressions allow an easier construction using logic scales.

9

Demodulations

9.1. Coherent demodulation

In the case of the coherent demodulation, we compare the phase of the received signal to the reference phase. The reference phase must be identical to that of the emission. It is given by a voltage-controlled oscillator (VCO). The phase control of this VCO comprises a phase feedback loop that allows us to close phase φ_0 and close frequency ω_0.

9.1.1. Coherent demodulation of order 2

The modulated received signal $s(t)$ is of order 2:

$$s(t) = V\cos(\omega_0 t + \varphi_0 + \phi) \text{ with } \phi = 0 \text{ or } \pi$$

The VCO is perfect and the reference signal is of the form:

$$\psi(t) = V'\cos(\omega_0 t + \varphi_0)$$

Figure 9.1 shows the product of the two signals. In fact, we can consider that we made a second modulation on reception.

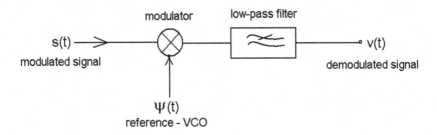

Figure 9.1. *Coherent demodulator of order 2*

Then the modulator on reception makes the product $v(t) = s(t) \cdot \psi(t)$:

$$v(t) = VV' \cos(\omega_0 t + \varphi_0 + \phi) \cdot \cos(\omega_0 t + \varphi_0)$$

That is:

$$v(t) = \frac{VV'}{2} \cos(2\omega_0 t + 2\varphi_0 + \phi) + \frac{VV'}{2} \cos\phi$$

Using the low-pass filter, we eliminate the part of $v(t)$ that oscillates at the double frequency $2\omega_0$, and:

$$\boxed{v(t) = v = \frac{VV'}{2} \cos\phi}$$

This means that when:

$$
\boxed{
\begin{array}{lll}
\phi = 0 & v = +\dfrac{VV'}{2} & \rightarrow \quad bit\ "0" \\[3mm]
\phi = \pi & v = -\dfrac{VV'}{2} & \rightarrow \quad bit\ "1"
\end{array}
}
$$

then it is possible to recover the two values of the previous binary signal.

9.1.2. *Coherent demodulation of order 4*

We recall that a coherent modulation of order 4 is in fact the sum of two coherent modulations of order 2 using two square carriers.

The coherent demodulation will also be made using two square carriers (Figure 9.2).

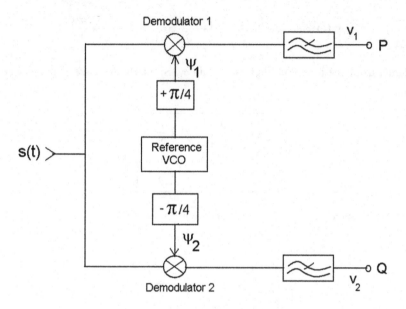

Figure 9.2. *Coherent demodulator of order 4*

The received modulated coherent signal is:

$$s(t) = V \cos\left(\omega_0 t + k\frac{\pi}{2} \right) \text{ with } k = 0, 1, 2, 3$$

and the reference VCO signals are:

$$\psi_1 = V' \cos\left(\omega_0 t + \frac{\pi}{4} \right)$$

$$\psi_2 = V' \cos\left(\omega_0 t - \frac{\pi}{4} \right)$$

After the demodulator 1, we obtain:

$$v_1(t) = V\cos\left(\omega_0 t + k\frac{\pi}{2}\right)\cdot V'\cos\left(\omega_0 t + \frac{\pi}{4}\right)$$

That is:

$$v_1(t) = \frac{VV'}{2}\cos\left(k\frac{\pi}{2} - \frac{\pi}{4}\right) + \frac{VV'}{2}\cos\left(2\omega_0 t + k\frac{\pi}{2} + \frac{\pi}{4}\right)$$

and after using the low-pass filter, we obtain v_1. Similarly, we have v_2 in the second branch:

$$v_1(t) = v_1 = \frac{VV'}{2}\cos\left(k\frac{\pi}{2} - \frac{\pi}{4}\right)$$

$$v_2(t) = v_2 = \frac{VV'}{2}\cos\left(k\frac{\pi}{2} + \frac{\pi}{4}\right)$$

With the values of $k = 0, 1, 2, 3$, we obtain v_1 and v_2:

k	v_1	v_2
0	$\frac{VV'}{2\sqrt{2}}$	$\frac{VV'}{2\sqrt{2}}$
1	$\frac{VV'}{2\sqrt{2}}$	$-\frac{VV'}{2\sqrt{2}}$
2	$-\frac{VV'}{2\sqrt{2}}$	$-\frac{VV'}{2\sqrt{2}}$
3	$-\frac{VV'}{2\sqrt{2}}$	$\frac{VV'}{2\sqrt{2}}$

We now have a negative logic using the correspondences:

$$+\frac{VV'}{2\sqrt{2}} \;\rightarrow\; bit\;"0"$$

$$-\frac{VV'}{2\sqrt{2}} \;\rightarrow\; bit\;"1"$$

and we obtain:

k	P	Q
0	0	0
1	0	1
2	1	1
3	1	0

The configuration of the demodulated signal is identical to that of the modulated signal:

$$\boxed{\begin{aligned} P \equiv A \\ Q \equiv B \end{aligned}}$$

9.2. Differential demodulation

Two carrier phases received during two intervals of time are compared with each other. This leads to extraction of the bound of phase and restoration of the signal.

9.2.1. Differential demodulation of order 2

We give the method of the demodulation in Figure 9.3.

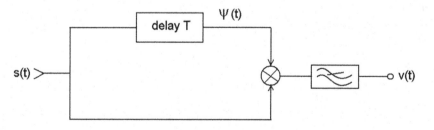

Figure 9.3. *Differential demodulator of order 2*

A part of the signal $s(t)$ is delayed by duration T to obtain $\psi(t)$:

$$s(t) = V\cos(\omega_0 t + \gamma)$$
$$\psi(t) = V'\cos(\omega_0(t - T) + \gamma)$$

where γ is an unimportant phase. However, in fact:

$$s(t + T) = V\cos(\omega_0 t + \gamma + \Delta\phi)$$

where $\Delta\phi$ is the bound of phase between the nth and the $(n+1)$th part.

We make the product:

$$v(t) = s(t + T)\cdot\psi(t) = VV'\cos(\omega_0(t - T) + \gamma)\cdot\cos(\omega_0 t + \gamma + \Delta\phi)$$

or:

$$v(t) = \frac{VV'}{2}\cos(\omega_0(2t - T) + 2\gamma + \Delta\phi) + \frac{VV'}{2}\cos(\omega_0 T + \Delta\phi)$$

and after using the low-pass filter:

$$\boxed{v(t) = v = \frac{VV'}{2}\cos(\Delta\phi + \omega_0 T)}$$

If we take:

$$\boxed{\omega_0 T = k\pi}$$

this gives:

$$\boxed{T = \frac{k}{2}T_0}$$

Now we have to consider the case where k is even and the case where k is odd.

If k is even:

$$\boxed{v = \frac{VV'}{2}\cos\Delta\phi}$$

and making the correspondence:

$$
\begin{array}{l}
v = +\dfrac{VV'}{2} \quad \rightarrow \quad bit \ "0" \\[2mm]
v = -\dfrac{VV'}{2} \quad \rightarrow \quad bit \ "1"
\end{array}
$$

we obtain:

v	$\Delta\phi$
0	0
1	π

to be compared with:

A	$\Delta\phi$
0	0
1	π

It is possible to recover the two values of the previous binary signal because $v = A$.

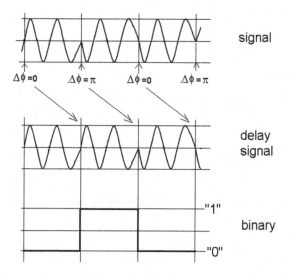

Figure 9.4. *Delay signal and the binary value*

In the other case when k is odd:

$$\boxed{v = -\frac{VV'}{2}\cos\Delta\phi}$$

We obtain:

v	$\Delta\phi$
-1	0
0	π

It is also possible to recover the two values of the previous binary signal because in this case we have $\boxed{v = \overline{A}}$.

9.2.2. *Differential demodulation of order 4*

We use the same principle as in the case of a differential modulation of order 2 (Figure 9.5).

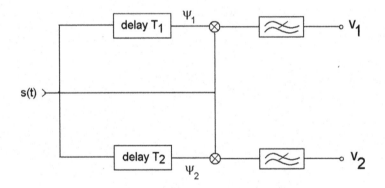

Figure 9.5. *Differential demodulator of order 4*

Consider the first arm, we have to assume the received signal and the delayed signal:

$$s(t) = V\cos(\omega_0 t + \gamma)$$
$$\psi_1(t) = V'\cos(\omega_0(t - T_1) + \gamma)$$

However, on multiplication, we have:

$$s(t+T) = V\cos(\omega_0 t + \gamma + \Delta\phi)$$

and after using the low-pass filter at v_1, we obtain:

$$v_1(t) = v_1 = \frac{VV'}{2}\cos(\Delta\phi + \omega_0 T_1)$$

We do the same on the arm v_2 and we obtain:

$$v_2(t) = v_2 = \frac{VV'}{2}\cos(\Delta\phi + \omega_0 T_2)$$

If we take:

$$\begin{cases} T_1 = (2k+1)\frac{T_0}{8} \text{ or } \omega_0 T_1 = (2k+1)\frac{\pi}{4} \\ T_2 = (2k+3)\frac{T_0}{8} \text{ or } \omega_0 T_2 = (2k+3)\frac{\pi}{4} \end{cases}$$

then we obtain:

$$\begin{cases} v_1 = \frac{VV'}{2}\cos\left(\Delta\phi + (2k+1)\frac{\pi}{4}\right) \\ v_2 = \frac{VV'}{2}\cos\left(\Delta\phi + (2k+3)\frac{\pi}{4}\right) \end{cases}$$

We recall that we want to recover:

$\Delta\phi$	A	B
0	0	0
$\pi/2$	0	1
π	1	1
$3\pi/2$	1	0

This is possible in the cases:

– for $k = 0$, we have $A = v_2$ and $B = v_1$;

– for $k = 1$, we have $A = v_1$ and $B = v_2$;

– for $k = 2$, we have $A = -v_2$ and $B = v_1$.

Then it is possible to recover the values of the previous binary signals.

Quality of the Digital Transmissions

10.1. Introduction

We will discuss the spectral power density in the cases of the coherent and differential modulations. We will also discuss the effect of the noise.

10.2. Differential demodulation of order 2

The carrier of frequency f_0 is modulated by an arbitrary numerical signal $s(t)$:

$$s(t) = V \cos(\omega_0 t + \gamma + \Delta\phi)$$

where:

$$\Delta\phi = 0 \ or \ \pi \ \text{ and } \ t \in \left[nT, (n+1)T\right]$$

γ : random phase equal distibuted on $[0, 2\pi]$

We show that power spectral density is:

$$\boxed{\Gamma(f) = \frac{V^2 T}{4}\left[\delta(f - f_0) + \delta(f + f_0)\right] * \left(\frac{\sin \pi fT}{\pi fT}\right)^2}$$

The second part of $\Gamma(f)$ is the power spectrum of the modulating square. Then, $\Gamma(f)$ is obtained by the two translations f_0 and $-f_0$ of the modulating square.

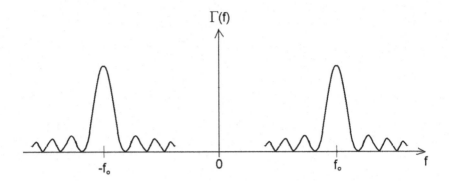

Figure 10.1. *Power spectral density of the differential demodulation of order 2*

A non-coherent modulation is equivalent to an amplitude modulation without carrier.

10.3. Coherent modulation of order 2

The signal of the two symbols is perfectly defined as:

$$s_1(t) = V \cos(\omega_0 t + 0 + \varphi_0)$$

$$s_2(t) = V \cos(\omega_0 t + \pi + \varphi_0)$$

where ω_0 is the pulsation of the carrier. We have a coherent signal in the case where $\omega_0 T = 2k\pi$ and k is an entire number.

The spectral density is:

$$\Gamma(f) = \frac{V^2 T}{4} \left[\delta(f - f_0) + \delta(f + f_0) \right] * \left(\frac{\sin \pi f T}{\pi f T} \right)^2$$
$$- \frac{V^2 T}{4} \left[\delta(f - f_0) - \delta(f + f_0) \right] * \left(\frac{1}{\pi T f_0} \cdot \frac{\sin^2 \pi f T}{\pi f T} \right)$$

The first term is the same as that in the case of a non-coherent modulation. We give the aspect of the density power only around f_0.

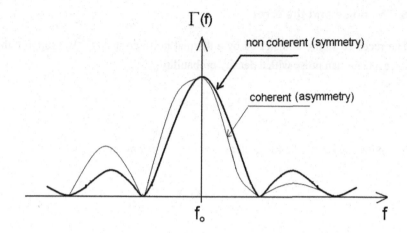

Figure 10.2. *Power spectral density of the coherent demodulation of order 2*

When $f_0 T$ is an entire number, we can write:

$$\Gamma(f) = \frac{V^2}{T} \cdot \frac{f_0^2 \sin^2 \pi f T}{\pi^2 \left(f^2 - f_0^2\right)^2}$$

We can see that the decrease at infinity of the spectra is $1/f^4$. In the case of a non-coherent modulation, it was only $1/f^2$.

10.4. Modulation (coherent or non-coherent) of order 4

If the modulation is coherent or non-coherent, the spectral is of the form:

$$\Gamma(f) = \frac{V^2 T}{4} \left[\delta(f - f_0) + \delta(f + f_0)\right] * \left(\frac{\sin \pi f T}{\pi f T}\right)^2$$

The coherence term has vanished.

For the same duration of the binary element, we have the same result as in the case of a two-state modulation.

This means that the spectral traffic is the same in PSK-2 and in PSK-4: but the quantity of information is double in the case of PSK-4.

10.5. The noise and the error

The received signal is distorted by a thermal noise around f_0. We suppose that we have a Gaussian noise with a density probability:

$$p(x) = \frac{1}{\sigma\sqrt{2\pi}} e^{-x^2/2\sigma^2}$$

The power noise (or the two-order moment) is the constant σ^2 because:

$$\overline{E}^2 = \int_{-\infty}^{+\infty} x^2 p(x)\,dx = \sigma^2$$

10.5.1. Modulation of order 2 of the phase

Due to the noise, the extremity of the signal vector is now not on the axis $0x$ (Figure 10.3). We have to add the noise vector $\vec{n}$.

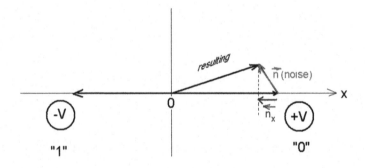

Figure 10.3. *Effect of the noise on the modulation of order 2. For a color version of the figure, see www.iste.co.uk/jarry/communications.zip*

Now we suppose that the two states "0" and "1" have the same probability.

If "0" is emitted, there will be an error on reception if:

$$\vec{n}_x \leq -V$$

which is the same as:

$$-\vec{n}_x \geq V$$

In addition, the error probability for "0" is:

$$\boxed{\mathcal{P}_e\left\{ \text{error if we have a "0"} \right\} = \mathcal{P}_e\left\{ \vec{n}_x \leq -V \right\}}$$

In the same way we have for "1":

$$\boxed{\mathcal{P}_e\left\{ \text{error if we have a "1"} \right\} = \mathcal{P}_e\left\{ \vec{n}_x \geq V \right\}}$$

These two probabilities are identical and it is sufficient to compute one of them, for example:

$$\mathcal{P}_e\left\{ \text{error "1"} \right\} = \frac{1}{\sigma\sqrt{2\pi}} \int_V^\infty e^{-\frac{x^2}{2\sigma^2}} dx$$

This can be written as:

$$\mathcal{P}_e\left\{ \text{error "1"} \right\} = \frac{1}{\sigma\sqrt{2\pi}} \left[\int_0^\infty e^{-\frac{x^2}{2\sigma^2}} dx - \int_0^V e^{-\frac{x^2}{2\sigma^2}} dx \right]$$

The first part is half of the Gauss integral:

$$\frac{1}{\sigma\sqrt{2\pi}} \int_{-\infty}^{+\infty} e^{-\frac{x^2}{2\sigma^2}} dx = 1$$

and:

$$\mathcal{P}_e\left\{ \text{error "1"} \right\} = \frac{1}{2} - \frac{1}{\sigma\sqrt{2\pi}} \int_0^V e^{-\frac{x^2}{2\sigma^2}} dx$$

We make the transformations:

$$\begin{cases} t = \dfrac{x}{\sigma\sqrt{2}} \\ erfX = \dfrac{2}{\sqrt{\pi}} \displaystyle\int_0^X e^{-t^2}\, dt; \quad \text{error function} \end{cases}$$

The two errors $\mathcal{P}_e\{\text{ error "0"}\} = \mathcal{P}_e\{\text{ error "1"}\}$ are identical then:

$$\boxed{\mathcal{P}_e = \frac{1}{2}\left[1 - erf\frac{V}{\sigma\sqrt{2}}\right]}$$

We also define the noise power:

$$N = \sigma^2$$

and the power of carrier:

$$C = \frac{V^2}{2}$$

We obtain:

$$\frac{C}{N} = \frac{V^2}{2\sigma^2}$$

or in decibels:

$$\boxed{\left(\frac{C}{N}\right)_{dB} = 10\log\frac{V^2}{2\sigma^2}}$$

Then the error probability on the symbols in the case of a modulation of order 2 of the phase will be:

$$\boxed{\mathcal{P}_e = \frac{1}{2}\left[1 - erf\sqrt{\frac{C}{N}}\right]}$$

10.5.2. Modulation of order 4 of the phase

S is the figurative point of the noisy signal. D_0, D_1, D_2, D_3 are the four decision regions limited by the axes φ_1 and φ_2.

There will be an error on the emitted symbol if S is not placed in a region that does not corresponds to the emitted phase.

Remember that to each symbol corresponds a binary element ("00", "01", "11", "10").

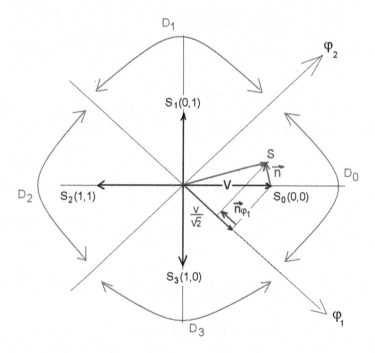

Figure 10.4. *Effect of the noise on the modulation of order 4. For a color version of the figure, see www.iste.co.uk/jarry/communications.zip*

For example, if S_0 has been emitted, then there is:

– 1 binary error if S belongs to the regions D_1 or D_3;

– 2 binary errors if S belongs to the region D_2.

If S_0 has been emitted, then the error is:

$$\mathcal{P}_e\{ \text{ error on } S_0\} = \frac{1}{2}\Big[\mathcal{P}_e\{ S \in D_1\} + \mathcal{P}_e\{ S \in D_3\} + 2\mathcal{P}_e\{ S \in D_2\}\Big]$$

This gives, for symmetry reasons:

$$\mathcal{P}_e\{ \text{ error on } S_0\} = \mathcal{P}_e\{ S \in D_1\} + \mathcal{P}_e\{ S \in D_2\}$$

which means:

$$\boxed{\mathcal{P}_e\{ \text{ error on } S_0\} = \mathcal{P}_e\{ S \in D_1 \cup D_2\}}$$

or:

$$\boxed{\mathcal{P}_e\{ \text{error on } S_0\} = \mathcal{P}_e\{ S \in D_2 \cup D_3\}}$$

Then we have an error when the point S belongs to the half plane (D_1, D_2) or (D_2, D_3).

This means that we have an error when the component of the noise on the axis φ_1, $\vec{n}_{\varphi_1}$, is lower than the quantity $-V/\sqrt{2}$.

Then:

$$\boxed{\mathcal{P}_e = \mathcal{P}_e\left\{ \vec{n}_{\varphi_1} \le -\frac{V}{\sqrt{2}}\right\}}$$

We recover the precedent result of a phase modulation of order 2 in which we made the transformation:

$$V \to \frac{V}{\sqrt{2}}$$

Then:

$$\boxed{\mathcal{P}_e = \frac{1}{2}\left[1 - erf\frac{V}{2\sigma}\right]}$$

The figure is preserved by rotation. Moreover, the same argument applies if we have emitted S_1, S_2 or S_3. In addition, the four states ("00", "01", "11", "10") have the same probability and the error will be given by:

$$\mathcal{P}_e = \frac{1}{2}\left[1 - erf\frac{V}{2\sigma}\right]$$

The error modulation of order 4 (phase) and of amplitude V is the same as that of order 2 but of amplitude $V/\sqrt{2}$ and gives the same noise.

This means that with constant noise the modulation of order 4 requires an amplitude $\sqrt{2}$ times larger than that of the modulation of order 2. The two modulations give the same errors and the same noise but require double the power.

Part 4

Exercises and Problems

11

Supports

11.1. Power links

We consider a link between two space stations. These space stations are characterized by:

– their distance R;

– the frequency of the transmitter f and the corresponding wavelength l;

– the power of the transmitter P_E and the emitting gain antenna G_E;

– the reception power P_R and the reception gain antenna G_R.

PROBLEM.–

1) Give the P_R / P_E ratio and deduce the reception power P_R.

2) Compute P_R (dBm and mW) in the case of a link of the satellite ATS-6 (Applications Technology Satellite-6) characterized by:

– an emitting power $P_E = 2$ Watts;

– an emitting gain $G_E = 37 \, \text{dB}$;

– a utilization frequency $f = 20 \, \text{GHz}$;

– a reception gain antenna $G_R = 45.8 \, \text{dB}$;

– a distance to the satellite $R = 36.941 \, \text{km}$.

SOLUTION.–

1) From the Friis formula, we have:

$$\boxed{\frac{P_R}{P_E} = \frac{G_R G_E \lambda^2}{(4\pi R)^2}}$$

This equation was given in the case of radar but it also occurs in the case of satellites.

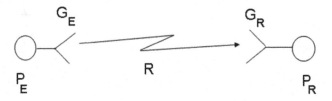

Figure 11.1. *Separation of two satellites*

Then:

$$P_R = \frac{G_R G_E \lambda^2}{(4\pi R)^2} P_E$$

Remember we have by definition:

$$\begin{cases} G_{RdB} = 10\log G_R \\ G_{EdB} = 10\log G_E \end{cases}$$

The same definitions occur in the case of the power:

$$\begin{cases} P_{RdB} = 10\log P_R \\ P_{EdB} = 10\log P_E \end{cases}$$

However, the definitions of the dBm are as follows:

$$\begin{cases} P_{RdBm} = 10\log\dfrac{P_R}{10^{-3}} \\[2mm] P_{EdBm} = 10\log\dfrac{P_E}{10^{-3}} \end{cases}$$

P is given in Watt(s) and we have the correspondence.

$P\,(Watts)$	$P\,(dBm)$
10^{-3}	0
10^{-2}	10
10^{-1}	20
1	30
etc.	etc.

Then:

$$\frac{P_R}{10^{-3}} = \frac{G_R G_E \lambda^2}{\left(4\pi R\right)^2} \frac{P_E}{10^{-3}}$$

and:

$$P_{RdBm} = 10\log G_R + 10\log G_E + 20\log \lambda - 20\log\left(4\pi R\right) + P_{EdBm}$$

2) Now we compute the wavelength:

$$\lambda = \frac{C}{f} = \frac{300\,000\ km/s}{20\ GHz} = \frac{300\,000 \cdot 10^3}{20 \cdot 10^9} = 1,5 \cdot 10^{-2}\,m = 1,5\ cm$$

and:

$$P_{RdBm} = 45.8 + 37 + 20\log\left(1.5 \cdot 10^{-2}\right) - 20\log\left(4\pi \cdot 36.941 \cdot 10^{+3}\right) + 10\log\left(2 \cdot 10^{+3}\right)$$

This gives:

$$P_{RdBm} = -34\ dBm$$

This means:

$$P_{RdBm} = 10\log\frac{P_R}{10^{-3}} \quad \text{or} \quad \frac{P_R}{10^{-3}} = 10^{P_{RdBm}/10}$$

$$P_R\left(Watts\right) = 10^{-3} \cdot 10^{P_{RdBm}/10}$$

and we have a very weak value:

$$\begin{cases} P_R \left(Watts \right) = 3.98 \cdot 10^{-7} \, Watts \\ P_R = 3.98 \cdot 10^{-4} \, mWatts \\ P_R = 0.398 \, \mu Watts \end{cases}$$

This can be obtained through the Friis formula by using the two gains (reception and emitting):

$$\begin{cases} G_{RdB} = 10 \log G_R \\ G_{EdB} = 10 \log G_E \end{cases}$$

This means:

$$\begin{cases} G_R = 10^{G_{RdB} \cdot 10} = 10^{4.58} = 38019 \\ G_E = 10^{G_{EdB} \cdot 10} = 10^{3.7} = 5012 \end{cases}$$

and using:

$$P_R = \frac{G_R G_E \lambda^2}{\left(4\pi R \right)^2} P_E$$

By this method, we obtain the same results:

$$\begin{cases} P_R \left(Watts \right) = 3.979 \cdot 10^{-7} \, Watts \\ P_R = 3.979 \cdot 10^{-4} \, mWatts \\ P_R = 0.3979 \, \mu Watts \end{cases}$$

11.2. Antennas

We consider two localized, punctual and isotropic antennas (Figure 11.2). Their excitations are defined by the equations:

– antenna no. 1: $E_1 = A e^{j\omega t}$;

– antenna no. 2: $E_2 = A e^{j(\omega t + \varphi)}$.

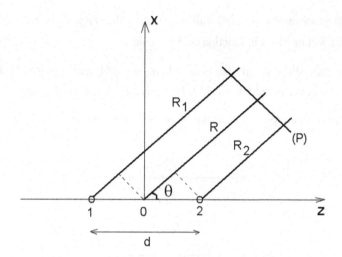

Figure 11.2. *Elementary antenna formed with two localized and punctual charges*

PROBLEM.–

1) Compute the total electric wave E_T in the plane (P) at the distance R from the origin 0 and in the direction θ. The distance between the two sources is d.

The waves coming from charges 1 and 2 arrive on the plane (P) at the instants $(t - R_1/C)$ and $(t - R_2/C)$, respectively, where C is the propagation speed in the environment (air).

Then compute R_1 and R_2 in function of R, d, θ and show that we have:

$$E_T = E_{MAX} \cos\left(k_0 \frac{d}{2} \cos\theta + \frac{\varphi}{2} \right)$$

where k_0 is the propagation constant:

$$k_0 = \frac{\omega}{C} = \frac{2\pi f}{C} = \frac{2\pi}{\lambda} T$$

With the wavelength λ and the period T, we have $C = \lambda/T$ and $fT = 1$.

We will give the expression of E_{MAX}.

2) In the case where $d = \lambda/2$ and $\varphi = 0$, give the diagram beam in the plane $\langle x0y \rangle$. This will be given in function of E_{MAX} and θ.

3) Give this diagram in the case where $d = \lambda/4$ and $\varphi = -\pi/2$. We will consider the successive angles $\theta = 0, \pi/2, \pi, 3\pi/2$. Give the conclusions.

4) We have four isotropic antennas to realize a network of antennas as shown in Figure 11.3.

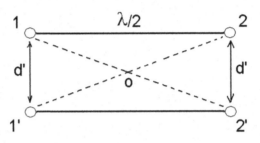

Figure 11.3. *A system of four antennas*

Antennas 1 and 2 have the same phase excitation. Antennas 1' and 2' also have the same phase excitation. Compute the phase φ' and the distance d' between the doublets (1,2) and (1',2'), which give the unidirectional system.

5) What are the conditions on φ' and d' that give this system as directional as possible? What is the expression of the total electric wave? Give the corresponding diagram beam.

SOLUTION.–

1) The waves coming from sources 1 and 2 arrive on the plane (P) at the moments $(t - R_1/C)$ and $(t - R_2/C)$, respectively. The total electric wave is:

$$E_T = Ae^{j\omega\left(t - \frac{R_1}{C}\right)} + Ae^{j\left[\omega\left(t - \frac{R_2}{C}\right) + \varphi\right]}$$

However

$$R_1 = R + \frac{d}{2}\cos\theta$$

$$R_2 = R - \frac{d}{2}\cos\theta$$

and:

$$E_T = Ae^{j\omega t - j\frac{\omega}{C}\left(R + \frac{d}{2}\cos\theta\right)} + Ae^{j\omega t - j\frac{\omega}{C}\left(R - \frac{d}{2}\cos\theta\right)}e^{j\varphi}$$

It is:

$$E_T = Ae^{j\omega t - j\frac{\omega}{C}R}e^{j\frac{\varphi}{2}}\left[e^{-j\frac{\omega}{C}\frac{d}{2}\cos\theta}e^{-j\frac{\varphi}{2}} + e^{j\frac{\omega}{C}\frac{d}{2}\cos\theta}e^{j\frac{\varphi}{2}}\right]$$

However, the propagation constant is:

$$k_0 = \frac{\omega}{C} = \frac{2\pi f}{C} = \frac{2\pi f}{\lambda}T = \frac{2\pi}{\lambda} \quad \text{because} \quad C = \frac{\lambda}{T} \quad \text{and} \quad fT = 1$$

Using these properties:

$$E_T = Ae^{j\left[\omega\left(t - \frac{R}{C}\right) + \frac{\varphi}{2}\right]}\left[e^{-j\left(k_0\frac{d}{2}\cos\theta + \frac{\varphi}{2}\right)} + e^{j\left(k_0\frac{d}{2}\cos\theta + \frac{\varphi}{2}\right)}\right]$$

It is:

$$\begin{cases} E_T = E_{MAX}\cos\left(k_0\frac{d}{2}\cos\theta + \frac{\varphi}{2}\right) \\ E_{MAX} = 2Ae^{j\left[\omega\left(t - \frac{R}{C}\right) + \frac{\varphi}{2}\right]} \end{cases}$$

2) In the case where $d = \lambda/2$ and $\varphi = 0$, we have:

$$E_T = E_{MAX}\cos\left(\frac{\pi}{2}\cos\theta\right)$$

We consider the particular cases:

– $\theta = 0$, gives $E_T = 0$;

– $\theta = \dfrac{\pi}{2}$, gives $E_T = E_{MAX}$;

– the same reasoning occurs when $\theta = \pi, \dfrac{3\pi}{2}, 2\pi$.

We show in Figure 11.4 the corresponding diagram beam.

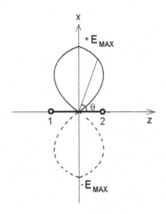

Figure 11.4. *Diagram with* $d = \lambda/2$ *and* $\varphi = 0$

3) In the case where $d = \lambda/4$ and $\varphi = -\pi/2$, we have:

$$\boxed{E_T = E_{MAX} \cos\frac{\pi}{4}(\cos\theta - 1)}$$

We consider the particular cases:

– $\theta = 0$, gives $E_T = E_{MAX}$ and the waves are in phase;

– $\theta = \dfrac{\pi}{2}$, gives $E_T = \dfrac{E_{MAX}}{\sqrt{2}}$;

– $\theta = \pi$, gives $E_T = 0$ and the waves are in opposite phase;

– the same reasoning occurs when $\theta = \dfrac{3\pi}{2}, 2\pi$.

We show in Figure 11.5 the corresponding diagram beam.

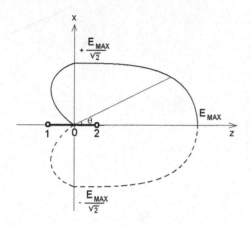

Figure 11.5. *Diagram with* $d = \lambda/4$ *and* $\phi = -\pi/2$

4) Sources 1 and 2 form an antenna $\lambda/2$. From the first question (no. 1), the wave is of the form:

$$Ae^{-j\omega\frac{R}{C}}e^{j\omega t}\cos\left(\frac{\pi}{2}\cos\theta\right)$$

This is:

$$E_M \cos\left(\frac{\pi}{2}\cos\theta\right)e^{j\omega t}$$

Sources 1' and 2' also form an antenna $\lambda/2$. By the same reasoning, we have a wave with a phase φ':

$$E_M \cos\left(\frac{\pi}{2}\cos\theta\right)e^{j(\omega t+\varphi')}$$

The total wave is the sum of the contributions of sources 1 and 2 and of sources 1' and 2':

$$E_T = E_M \cos\left(\frac{\pi}{2}\cos\theta\right)e^{j\omega\left(t-\frac{R'_1}{C}\right)} + E_M \cos\left(\frac{\pi}{2}\cos\theta\right)e^{j\omega\left(t-\frac{R'_2}{C}\right)}e^{j\varphi'}$$

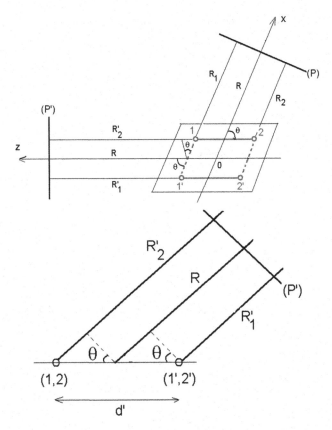

Figure 11.6. *Antenna formed with four localized and punctual charges. For a color version of the figure, see www.iste.co.uk/jarry/communications.zip*

Then we have in this case:

$$R'_1 = R - \frac{d'}{2}\sin\theta$$

$$R'_2 = R + \frac{d'}{2}\sin\theta$$

This gives:

$$E_T = E_M \cos\left(\frac{\pi}{2}\cos\theta\right) e^{j\omega\left(t - \frac{R}{C}\right)} \left[e^{j\frac{\omega}{C}\frac{d'}{2}\sin\theta} + e^{-j\frac{\omega}{C}\frac{d'}{2}\sin\theta} e^{j\varphi'} \right]$$

or:

$$E_T = 2E_M \cos\left(\frac{\pi}{2}\cos\theta\right)e^{j\omega\left(t-\frac{R}{C}\right)}e^{j\frac{\varphi}{2}}\cos\left[\frac{\omega\,d'}{2C}\sin\theta - \frac{\varphi'}{2}\right]$$

Using $\omega/C = 2\pi/\lambda$, we can write:

$$E_T = E_m \cos\left(\frac{\pi}{2}\cos\theta\right)\cdot\cos\left[\frac{\pi\,d'}{\lambda}\sin\theta - \frac{\varphi'}{2}\right]$$

5) The system is as directional as possible if:

a) $\dfrac{\pi\,d'}{\lambda}\sin\theta - \dfrac{\varphi'}{2} = 0$ when $\theta = \pi/2$, which corresponds to a maximum of E_T;

b) $\dfrac{\pi\,d'}{\lambda}\sin\theta - \dfrac{\varphi'}{2} = \pm\dfrac{\pi}{2}$ when $\theta = -\pi/2$, which corresponds to a minimum of E_T.

If we take for example the + solution in a), we obtain:

$$\begin{cases}\dfrac{\pi\,d'}{\lambda} - \dfrac{\varphi}{2} = 0 \\[2mm] -\dfrac{\pi\,d'}{\lambda} - \dfrac{\varphi}{2} = -\dfrac{\pi}{2}\end{cases}$$

The resolution of these two equations gives $\varphi' = \pi/2$ and $d' = \lambda/4$, and the total electric wave:

$$E_T = E_m \cos\left(\frac{\pi}{2}\cos\theta\right)\cdot\cos\frac{\pi}{4}(\sin\theta - 1)$$

We consider the particular cases:

– $\theta = 0$, gives $E_T = 0$;

– $\theta = \dfrac{\pi}{2}$, gives $E_T = E_m$;

– $\theta = \pi$, gives $E_T = 0$;

– $\theta = 3\pi/2$, also gives $E_T = 0$.

The diagram is non-symmetrical and, as shown in Figure 11.7, there are only two small lobes in the direction $(-z)$.

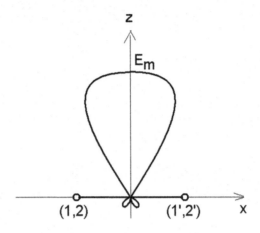

Figure 11.7. *Diagram of the four sources*

11.3. Simple designs of duplexers and multiplexers

We will give a simple design to realize duplexers and multiplexers. The design is made by the electric properties of a three port (Figure 11.8).

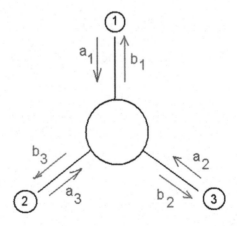

Figure 11.8. *Input and output waves of a three port. For a color version of the figure, see www.iste.co.uk/jarry/communications.zip*

PROBLEM.–

1) Give the scattering matrix (S) of the three ports.

2) Now we suppose that we can adapt simultaneously the three inputs. What is the form of this scattering matrix (S)?

3) The three arms are indistinguishable. How this matrix is formed?

4) We have the preservation of the energy if the junction is lossless and the matrix (S) is now a unitary matrix:

$$(S) \cdot (\overline{S})' = (I)$$

What are the solutions to have a unitary matrix (S)?

5) An arm of the three ports is closed on a dipole characterized by the reflection coefficient s_{11}. The system is now a four pole. Determine the new scattering matrix (Σ) of this system?

6) Give the possible applications as duplexer and multiplexer.

SOLUTION.–

1) The scattering matrix gives reflecting waves in function of the incident waves:

$$\begin{cases} b_1 = S_{11}a_1 + S_{12}a_2 + S_{13}a_3 \\ b_2 = S_{21}a_1 + S_{22}a_2 + S_{23}a_3 \\ b_3 = S_{31}a_1 + S_{32}a_2 + S_{33}a_3 \end{cases}$$

This is:

$$(b) = (S)(a)$$

2) We adapt simultaneously the three inputs:

$$S_{11} = S_{22} = S_{33} = 0$$

and:

$$(S) = \begin{pmatrix} 0 & S_{12} & S_{13} \\ S_{21} & 0 & S_{23} \\ S_{31} & S_{32} & 0 \end{pmatrix}$$

3) The three arms are indistinguishable:

$$\begin{cases} S_{12} = S_{23} = S_{31} = x \\ S_{21} = S_{13} = S_{32} = y \end{cases}$$

and:

$$(S) = \begin{pmatrix} 0 & x & y \\ y & 0 & x \\ x & y & 0 \end{pmatrix}$$

4) To get preservation of energy, the matrix (S) has to be unitary $(S)\cdot(\bar{S})' = (I)$:

$$\begin{pmatrix} 0 & x & y \\ y & 0 & x \\ x & y & 0 \end{pmatrix} \cdot \begin{pmatrix} 0 & \bar{y} & \bar{x} \\ \bar{x} & 0 & \bar{y} \\ \bar{y} & \bar{x} & 0 \end{pmatrix} = \begin{pmatrix} 1 & 0 & 0 \\ 0 & 1 & 0 \\ 0 & 0 & 1 \end{pmatrix}$$

This means:

$$\begin{cases} x\bar{y} = 0 \\ x\bar{x} + y\bar{y} = 1 \end{cases}$$

We have two solutions:

$$\begin{cases} x = 0 \\ y = e^{j\eta} \end{cases} \quad \text{or} \quad \begin{cases} x = e^{j\theta} \\ y = 0 \end{cases}$$

We choose the second solution, it gives:

$$\begin{cases} b_1 = e^{j\theta} a_2 \\ b_2 = e^{j\theta} a_3 \\ b_3 = e^{j\theta} a_1 \end{cases}$$

and the scattering matrix is:

$$(S) = e^{j\theta} \begin{pmatrix} 0 & 1 & 0 \\ 0 & 0 & 1 \\ 1 & 0 & 0 \end{pmatrix}$$

5) One port of the three ports (circulator) is closed on a dipole characterized by the reflection coefficient s_{11} (Figure 11.9).

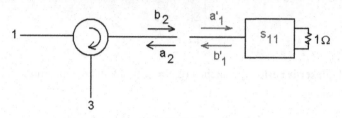

Figure 11.9. *A circulator closed on one port. For a color version of the figure, see www.iste.co.uk/jarry/communications.zip*

Then we have:

$$b'_1 = s_{11}a'_1$$

This means:

$$a_2 = s_{11}b_2$$

Reporting in the first equation of the second solution:

$$b_1 = s_{11}e^{j\theta}b_2$$

Using this result in the second and third equations:

$$\begin{cases} b_1 = s_{11}e^{2j\theta}a_3 \\ b_3 = e^{j\theta}a_1 \end{cases}$$

or:

$$\begin{pmatrix} b_1 \\ b_3 \end{pmatrix} = \begin{pmatrix} 0 & s_{11}e^{2j\theta} \\ e^{j\theta} & 0 \end{pmatrix} \cdot \begin{pmatrix} a_1 \\ a_3 \end{pmatrix}$$

This is a four-pole system and its new scattering matrix (Σ) is:

$$(\Sigma) = \begin{pmatrix} 0 & s_{11}e^{2j\theta} \\ e^{j\theta} & 0 \end{pmatrix}$$

6) This is the scattering matrix of a duplexer:

$$\begin{cases} \Sigma_{11} = \Sigma_{22} = 0 \\ \Sigma_{12} = s_{11}e^{2j\theta} \\ \Sigma_{21} = e^{j\theta} \end{cases}$$

This duplexer is perfectly matched ($\Sigma_{11} = \Sigma_{22} = 0$) and the magnitudes of Σ_{12} and s_{11} are the same:

$$\left| |\Sigma_{12}| = |s_{11}| \right|$$

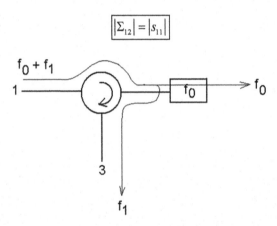

Figure 11.10. *Separation of two frequencies – duplexer*

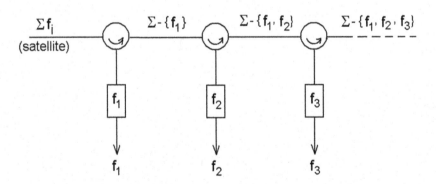

Figure 11.11. *Separation of several frequencies – multiplexer*

Depiction of a phase of the compartment of the system is the same as a reflection filter. As shown in Figure 11.10, it can be used to separate two frequencies (or two bands of frequencies). Near the filter and at a frequency different from f_0, there is a total reflection and f_1 comes on the port 3 of the circulator.

More generally, several circulators can be used to separate several channels centered at $f_1, f_2, f_3,...,f_i,...$ with the system shown in Figure 11.11. We have made a multiplexer.

Modulations and Demodulations Without Noise

12.1. Amplitude modulation system

We consider the ring modulator that allows us to modulate the numerical tension V_{MN} formed of "0" and "1" by a sinusoidal carrier of pulsation ω_C (Figure 12.1). We want to determine this modulated voltage V_S.

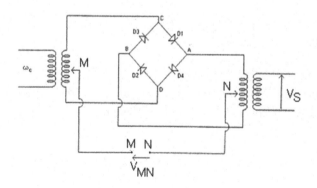

Figure 12.1. *Ring modulator*

PROBLEM.–

1) The points M and N are in the middle of the input transformer and also in the middle of the output transformer. The diodes D_1, D_2, D_3 and D_4 are considered

to be ideal and identical. The tensions V_M and V_N are not identical. Thus, what is the expression of V_S? Give the form of V_S. This represents the logic number "1".

2) We represent the logic number "0" when the tensions V_M and V_N are identical. In this case, give the form of V_S.

3) We propose to transmit the numerical logic number "1", "0", "1". Give the form of the output signal V_S.

4) In fact, this signal is filtered because the band of the system is limited. Then give the form of V_S?

5) In the case of the logical number "1" to be transmitted, the output signal is of the form $V_S = S(\omega_C) \cdot V_{MN}$:

a) What are the properties of the signal $S(\omega_C)$? And give the Fourier series of this signal.

b) The modulator "gives" only the first component of $S(\omega_C)$. Justify the results of question 4.

SOLUTION.–

1) Without loss of generality, we consider the case where $V_M \succ V_N$. We associate $V_{MN} \succ 0$ with a logical number "1".

a) During the positive alternate of the carrier (ω_C): $V_C \succ V_D$. The current goes along the points CAD because the diodes D_1 and D_4 are conductive and the diodes D_2 and D_3 are not conductive. Then from Figure 12.2 we have at the output $\boxed{V_S = V_{MN}}$.

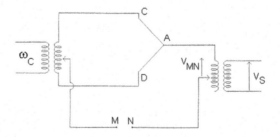

Figure 12.2. D_1 and D_4 conductive, D_2 and D_3 not conductive

b) In the case where $V_C \prec V_D$, the current goes along the points CBD. The diodes D_2 and D_3 are conductive and the diodes D_1 and D_4 are not conductive. Then from Figure 12.3, we have $\boxed{V_S = -V_{MN}}$.

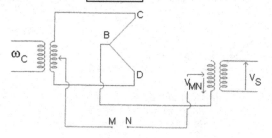

Figure 12.3. D_2 and D_3 conducting, D_1 and D_4 not conducting

We can now give the curve V_S in function of the time t (Figure 12.4).

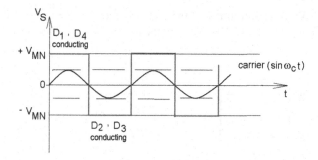

Figure 12.4. V_S in function of the time t (red curve) when (D_1, D_4) or (D_2, D_3) are conductive. For a color version of the figure, see www.iste.co.uk/jarry/communications.zip

We can also write:

$$\boxed{V_S = S(\omega_C) \cdot V_{MN}}$$

where $S(\omega_C)$ is a square signal of pulsation ω_C and of amplitude ± 1.

Then to the logic number "1" we make the correspondence $V_S = S(\omega_C) \cdot V_{MN}$:

$$\boxed{\text{"1"} \rightarrow V_S = S(\omega_C) \cdot V_{MN}}$$

2) Nothing passes because diodes D_1 and D_4 (and also D_2 and D_3) are at zero voltage (Figure 12.5).

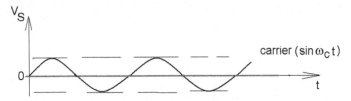

Figure 12.5. V_S *in function of the time t (red curve) when all the diodes are at a zero voltage. For a color version of the figure, see www.iste.co.uk/jarry/communications.zip*

and we have:

$$"0" \rightarrow V_S = 0$$

3) and 4) We wanted to transmit the numerical logic number "1", "0", "1". The forms of the output signal V_S are given by the next curves (Figure 12.6).

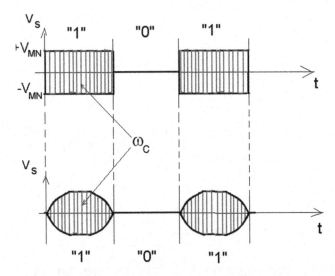

Figure 12.6. *Logic number "1","0","1" with a non-limited and a limited band. For a color version of the figure, see www.iste.co.uk/jarry/communications.zip*

5) $S(\omega_C)$ is a square signal of amplitude ± 1 and of pulsation ω_C (Figure 12.7).

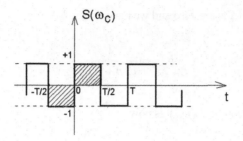

Figure 12.7. *Square signal* $S(\omega_c)$. *For a color version of the figure, see www.iste.co.uk/jarry/communications.zip*

For the development in Fourier series, we have to calculate the three components a_0, a_n and b_n:

$$\begin{cases} S(t) = a_0 + \sum a_n \cos n\omega t + \sum b_n \sin n\omega t \\ a_0 = \dfrac{1}{T}\int_0^T S(t)\,dt = \text{mean value and } \omega = 2\pi / T \\ a_n = \dfrac{2}{T}\int_0^T S(t)\cos n\omega t\,dt \\ b_n = \dfrac{2}{T}\int_0^T S(t)\sin n\omega t\,dt \end{cases}$$

$S(t)$ is an odd function then $a_0 = a_n = 0$. b_n is calculated on a period $[-T/2, +T/2]$:

$$b_n = \frac{2}{T}\int_{-T/2}^{+T/2} S(t)\sin\frac{2\pi nt}{T}\,dt = -\frac{2}{T}\int_{-T/2}^{0} \sin\frac{2\pi nt}{T}\,dt + \frac{2}{T}\int_{0}^{+T/2} \sin\frac{2\pi nt}{T}\,dt$$

The two surfaces between $[-T/2, 0]$ and $[0, +T/2]$ are equivalent:

$$b_n = \frac{4}{T}\int_{0}^{+T/2} \sin\frac{2\pi nt}{T}\,dt$$

This gives:

$$\boxed{b_n = \frac{2}{n\pi}\left(1 - \cos n\pi\right) = \frac{2}{n\pi}\left(1 - (-1)^n\right)}$$

The odd coefficients are zero and we have only:

$$b_{2p+1} = \frac{4}{(2p+1)\pi}$$

The decomposition of $S(\omega_C)$ is now:

$$S(\omega_C) = \sum_{p=0}^{\infty} \frac{4}{(2p+1)\pi} \sin(2p+1)\omega_C t$$

and:

$$V_S = S(\omega_C)V_{MN} = \frac{4V_{MN}}{\pi} \sum_{p=0}^{\infty} \frac{\sin(2p+1)\omega_C t}{(2p+1)}$$

By the filtering of the system, we do not consider the terms corresponding to $p \geq 1$ and there remains only the term $p = 0$ and:

$$V_S = \frac{4}{\pi}V_{MN} \sin \omega_C t$$

This result justifies question 4.

12.2. Coherent amplitude demodulation system

The following system allows us to demodulate $a(t)$ into a signal AM:

$$a(t) = A(1 + km(t))\cos(\omega_C t + \varphi)$$

The frequency carrier ω_C is very much greater than the maximum value of the carrier of the signal $m(t)$ to be transported.

k has a value so that:

$$-1 \leq km(t) \leq +1$$

We multiply the signal $a(t)$ by another signal $b(t)$:

$$b(t) = B\cos(\omega_c t + \varphi)$$

where $b(t)$ is coherent with $a(t)$. It has the same carrier ω_c and the same phase φ.

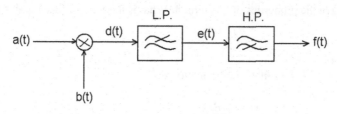

Figure 12.8. *Coherent demodulation of* AM

PROBLEM.–

1) Determine the signal $d(t)$ after multiplication.

2) This signal $d(t)$ is filtered by a low-pass (LP) filter. Determine the expression of $e(t)$ after this filter.

3) Then $e(t)$ is going through a high-pass (HP) filter, which is shown in Figure 12.9.

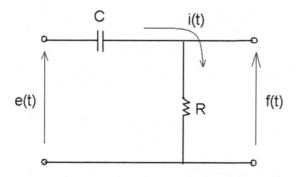

Figure 12.9. *The high-pass filter*

What is the transfer function of this RC filter?

$$H(j\omega) = \frac{F(j\omega)}{E(j\omega)}$$

4) Trace the modulus of this transfer function versus ω.

Then give the properties of this circuit. What is the form of the signal $f(t)$?

SOLUTION.–

1) The signal $d(t)$ after multiplication is:

$$d(t) = AB(1 + k\, m(t))\cos^2(\omega_c t + \varphi)$$

or:

$$d(t) = \frac{AB}{2}(1 + k\, m(t))\left[1 + \cos 2(\omega_c t + \varphi)\right]$$

2) After using the LP filter, we have:

$$\boxed{e(t) = \frac{AB}{2}(1 + k\, m(t))}$$

3) If $I(j\omega)$ is the current in the capacity C or in the resistance R, we have:

$$\begin{cases} E(j\omega) = \dfrac{I(j\omega)}{j\omega C} + F(j\omega) \\ F(j\omega) = R I(j\omega) \end{cases}$$

It gives:

$$E(j\omega) = \frac{F(j\omega)}{j\omega RC} + F(j\omega)$$

We can find the transfer function:

$$\boxed{H(j\omega) = \frac{F(j\omega)}{E(j\omega)} = \frac{j\omega RC}{1 + j\omega RC}}$$

and the modulus:

$$H(j\omega)\cdot H(-j\omega)=\frac{\omega^2\,R^2C^2}{1+\omega^2\,R^2C^2}$$

$$\boxed{\left|H(\omega)\right|=\frac{\omega\,RC}{\sqrt{1+\omega^2\,R^2C^2}}}$$

4) The graph is given by the curve shown in Figure 12.10.

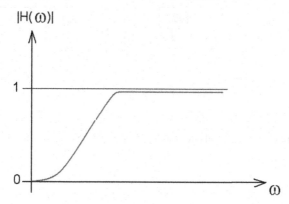

Figure 12.10. *Response of the high pass filter. For a color version of the figure, see www.iste.co.uk/jarry/communications.zip*

This is the graph of an HP filter because:

$$\begin{cases} H(0)=0 \\ H(\infty)=1 \end{cases}$$

The zero-frequency $(\omega(0)=0)$ component is naught and we have:

$$\boxed{f(t)=\frac{AB}{2}k\,m(t)}$$

The expression of $f(t)$ is the same as that of $m(t)$. It is only multiplied by the factor $\dfrac{AB}{2}k$. The signal $m(t)$ is then demodulated but this demodulation is possible by using a coherent signal.

12.3. Quadrature multiplexing system

PROBLEM.–

1) Let us consider the quadrature multiplexing system that allows us to transport the signals m_1 and m_2 using only one carrier ω_C (Figure 12.11).

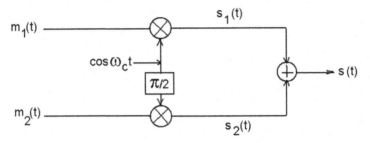

Figure 12.11. *Quadrature multiplex*

What are the expressions of the two signals $s_1(t)$ and $s_2(t)$ and also of their sum $s(t)$?

2) Demodulation is made on reception by using the carrier as shown in Figure 12.12.

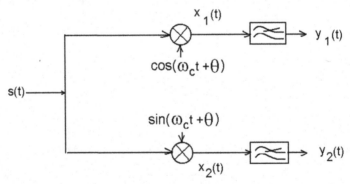

Figure 12.12. *Demodulation*

What are the expressions of the signals $x_1(t)$ and $x_2(t)$? What happens to these expressions when $\theta = 0$ and when θ is weak? Imagine a network to recover the signals m_1 and m_2.

3) Now emission uses only one coherent oscillator ω_c but it is followed by two filters (one LP and one HP) (Figure 12.13).

What are the expressions of the signals all along the emitting chain?

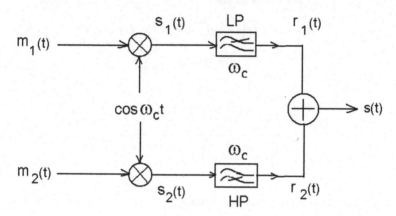

Figure 12.13. *Modulation using one coherent oscillator*

4) The demodulation is made using an envelope detection (Figure 12.14).

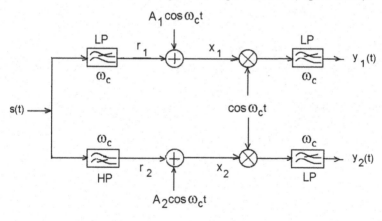

Figure 12.14. *Demodulation*

Give the signals $r_1(t)$ and $r_2(t)$ and also $x_1(t)$ and $x_2(t)$. What happens to these signals after the envelope detection?

SOLUTION.–

1) We have on reception:

$$\begin{cases} s_1(t) = m_1(t)\cos\omega_c t \\ s_2(t) = m_2(t)\cos\left(\omega_c t + \dfrac{\pi}{2}\right) = -m_2(t)\sin\omega_c t \end{cases}$$

and:

$$s(t) = s_1(t) + s_2(t)$$

$$\boxed{s(t) = m_1(t)\cos\omega_c t - m_2(t)\sin\omega_c t}$$

The problem is to separate the two signals $m_1(t)$ and $m_2(t)$ from their sum.

2) The demodulation will be made by using the carrier as shown in Figure 12.12. The reception oscillator is *synchronized* with the emission oscillator (the two have the same oscillation frequency ω_C but they can have a different phase θ).

Using the schema, we have:

$$\begin{cases} x_1(t) = s(t) \cdot \cos(\omega_c t + \theta) \\ x_1(t) = m_1(t)\cos\omega_c t \cdot \cos(\omega_c t + \theta) - m_2(t)\sin\omega_c t \cdot \cos(\omega_c t + \theta) \end{cases}$$

and using the following identities:

$$\begin{cases} 2\cos a \cdot \cos b = \cos(a+b) + \cos(a-b) \\ 2\sin a \cdot \cos b = \sin(a+b) + \sin(a-b) \\ 2\sin a \cdot \sin b = -\cos(a+b) + \cos(a-b) \end{cases}$$

we get:

$$\boxed{\begin{aligned} x_1(t) &= \frac{m_1(t)}{2}\cos(2\omega_c t + \theta) + \frac{m_1(t)}{2}\cos\theta \\ &\quad - \frac{m_2(t)}{2}\sin(2\omega_c t + \theta) + \frac{m_2(t)}{2}\sin\theta \end{aligned}}$$

We do the same for $x_2(t)$ and we have:

$$\begin{cases} x_2(t) = s(t) \cdot \sin(\omega_C t + \theta) \\ x_2(t) = m_1(t)\cos\omega_C t \cdot \sin(\omega_C t + \theta) - m_2(t)\sin\omega_C t \cdot \sin(\omega_C t + \theta) \end{cases}$$

and:

$$\boxed{\begin{aligned} x_2(t) &= \frac{m_1(t)}{2}\sin(2\omega_C t + \theta) + \frac{m_1(t)}{2}\sin\theta \\ &+ \frac{m_2(t)}{2}\cos(2\omega_C t + \theta) - \frac{m_2(t)}{2}\cos\theta \end{aligned}}$$

In the coherence case $\theta = 0$:

$$\boxed{\begin{cases} x_1(t) = \dfrac{m_1(t)}{2}\cos 2\omega_C t - \dfrac{m_2(t)}{2}\sin 2\omega_C t + \dfrac{m_1(t)}{2} \\ x_2(t) = \dfrac{m_1(t)}{2}\sin 2\omega_C t + \dfrac{m_2(t)}{2}\cos 2\omega_C t - \dfrac{m_2(t)}{2} \end{cases}}$$

In this case, the demodulation is very easy; it is sufficient to use an LP filter of cutoff ω_C (Figure 12.15).

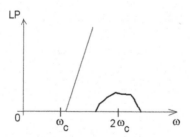

Figure 12.15. *The LP to recover $m_1(t)$ and $m_2(t)$. For a color version of the figure, see www.iste.co.uk/jarry/communications.zip*

In addition, we have:

$$\boxed{\begin{cases} y_1(t) = \dfrac{m_1(t)}{2} \\ y_2(t) = -\dfrac{m_2(t)}{2} \end{cases}}$$

If $\theta \neq 0$ and after the LP filters we have:

$$\begin{cases} y_1(t) = \dfrac{m_1(t)}{2}\cos\theta + \dfrac{m_2(t)}{2}\sin\theta \\ y_2(t) = \dfrac{m_1(t)}{2}\sin\theta - \dfrac{m_2(t)}{2}\cos\theta \end{cases}$$

If we also have a very weak value of θ, this means $\cos\theta \approx 1$ and $\sin\theta \approx \theta$:

$$\begin{cases} y_1(t) = \dfrac{m_1(t)}{2}\cos\theta\left\{1 + \dfrac{m_2}{m_1}tg\theta\right\} \\ y_2(t) = -\dfrac{m_2(t)}{2}\cos\theta\left\{1 - \dfrac{m_1}{m_2}tg\theta\right\} \end{cases}$$

and:

$$\begin{cases} y_1(t) \approx \dfrac{m_1(t)}{2}\left\{1 + \dfrac{m_2}{m_1}\theta\right\} \\ y_2(t) \approx -\dfrac{m_2(t)}{2}\left\{1 - \dfrac{m_1}{m_2}\theta\right\} \end{cases}$$

However, $m_1(t)$ and $m_2(t)$ are of the same order:

$$\begin{cases} y_1(t) \approx \dfrac{m_1(t)}{2}\{1 + \theta\} \\ y_2(t) \approx -\dfrac{m_2(t)}{2}\{1 - \theta\} \end{cases}$$

where θ is the interference signal, and if we want an interference maximum of 1% we have to take $\theta \approx 0.01 Rd$.

3) Now at the emission we use only one oscillator of frequency ω_C followed by one LP and one HP filter. Then after the two multiplications, we have:

$$\begin{cases} s_1(t) = m_1(t)\cos\omega_c t = \dfrac{m_1}{2}e^{j\omega_c t} + \dfrac{m_1}{2}e^{-j\omega_c t} \\ s_2(t) = m_2(t)\cos\omega_c t = \dfrac{m_2}{2}e^{j\omega_c t} + \dfrac{m_2}{2}e^{-j\omega_c t} \end{cases}$$

It gives in the frequency space by using the Fourier transform (FT):

$$s_1(t) \xrightarrow{FT} S_1(\omega) \;\; ; \;\; m_1(t) \xrightarrow{FT} M_1(\omega)$$
$$s_2(t) \xrightarrow{FT} S_2(\omega) \;\; ; \;\; m_2(t) \xrightarrow{FT} M_2(\omega)$$

$$\begin{cases} S_1(\omega) = \dfrac{1}{2}M_1(\omega - \omega_c) + \dfrac{1}{2}M_1(\omega + \omega_c) \\ S_2(\omega) = \dfrac{1}{2}M_2(\omega - \omega_c) + \dfrac{1}{2}M_2(\omega + \omega_c) \end{cases}$$

and, for example, after the LP filter in the case of $s_1(t)$ (Figure 12.16).

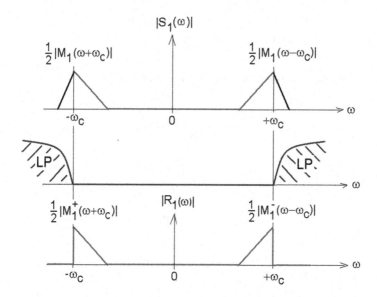

Figure 12.16. *Influence of the LP filter to* $s_1(t)$. *For a color version of the figure, see www.iste.co.uk/jarry/communications.zip*

We obtain an inferior single side band:

$$R_1(\omega) = \frac{1}{2} M_1^- (\omega - \omega_C) + \frac{1}{2} M_1^+ (\omega + \omega_C)$$

By the same way, we get for the signal $s_2(t)$ a superior single side band:

$$R_2(\omega) = \frac{1}{2} M_2^+ (\omega - \omega_C) + \frac{1}{2} M_2^- (\omega + \omega_C)$$

The two spectra do not overlap then it is possible to transmit the two signals with the same system, we have made a duplexer (Figure 12.17).

It is shown that coming back in the time domain that:

$$\begin{cases} r_1(t) = \dfrac{m_1}{2} \cos \omega_C t + \dfrac{\hat{m}_1}{2} \sin \omega_C t \\[2mm] r_2(t) = \dfrac{m_2}{2} \cos \omega_C t - \dfrac{\hat{m}_2}{2} \sin \omega_C t \end{cases}$$

where $\hat{m}(t)$ is the Hilbert transform of $m(t)$.

In addition, we have for the two signals $m_1(t)$ and $m_2(t)$.

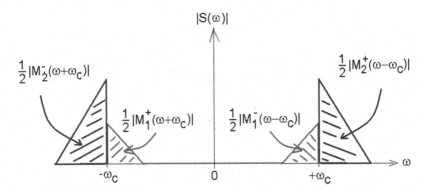

Figure 12.17. *The two FT do not overlap. For a color version of the figure, see www.iste.co.uk/jarry/communications.zip*

In addition:

$$S(\omega) = R_1(\omega) + R_2(\omega)$$

and:

$$s(t) = r_1(t) + r_2(t)$$

4) For demodulation, we superpose to the two signals $r_1(t)$ and $r_2(t)$ the two signals $A_1 \cos \omega_c t$ and $A_2 \cos \omega_c t$. It gives:

$$\begin{cases} x_1(t) = \left(A_1 + \dfrac{m_1}{2} \right) \cos \omega_c t + \dfrac{\hat{m}_1}{2} \sin \omega_c t \\[4mm] x_2(t) = \left(A_2 + \dfrac{m_2}{2} \right) \cos \omega_c t - \dfrac{\hat{m}_2}{2} \sin \omega_c t \end{cases}$$

With the two coefficients A_1 and A_2 as large as:

$$\begin{cases} A_1 + \dfrac{m_1}{2} \gg \dfrac{\hat{m}_1}{2} \\[4mm] A_2 + \dfrac{m_2}{2} \gg \dfrac{\hat{m}_2}{2} \end{cases}$$

and:

$$\begin{cases} x_1(t) \approx \left(A_1 + \dfrac{m_1}{2} \right) \cos \omega_c t \\[4mm] x_2(t) \approx \left(A_2 + \dfrac{m_2}{2} \right) \cos \omega_c t \end{cases}$$

These two signals are now multiplied by $\cos \omega_c t$ and we obtain:

$$\begin{cases} x_1 \cos \omega_c t = \left(A_1 + \dfrac{m_1}{2} \right) \cos^2 \omega_c t = \left(\dfrac{A_1}{2} + \dfrac{m_1}{4} \right) [1 + \cos 2\omega_c t] \\[4mm] x_2 \cos \omega_c t = \left(A_2 + \dfrac{m_2}{2} \right) \cos^2 \omega_c t = \left(\dfrac{A_2}{2} + \dfrac{m_2}{4} \right) [1 + \cos 2\omega_c t] \end{cases}$$

And after the two LP filters, we have:

$$\begin{cases} y_1(t) = \left(\dfrac{A_1}{2} + \dfrac{m_1}{4} \right) \\ y_2(t) = \left(\dfrac{A_2}{2} + \dfrac{m_2}{4} \right) \end{cases}$$

An elimination of the two constants A_1 and A_2 gives us the two signals $m_1(t)$ and $m_2(t)$:

$$\begin{cases} z_1(t) \approx \dfrac{m_1}{4} \\ z_2(t) \approx \dfrac{m_2}{4} \end{cases}$$

13

Modulations and Demodulations With Noise

13.1. Demodulation in the presence of noise using a triangular filter

We want to transmit a binary and noisy signal $x(t)$ that is written in a simple form as:

$$x(t) = \sum_{k=-\infty}^{+\infty} a_k \delta(t - kT) + n(t)$$

– $x(t)$ is a series of Dirac impulses $\delta(t)$;

– weighted by the binary digit $a_k = \pm 1$;

– $n(t)$ is a white Gaussian noise with a bilateral density power $N_0 / 2$.

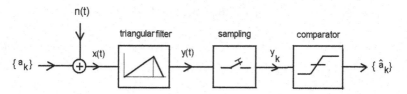

Figure 13.1. *Reception system using a triangular filter*

As shown in Figure 13.2, the reception filter has an impulse triangular response $h(t)$ and we give:

$$\int_{-\infty}^{+\infty} |h(t)|^2 = \frac{4}{3T_0}$$

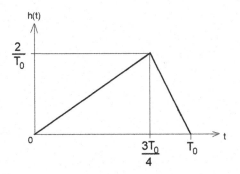

Figure 13.2. *Triangular impulse response of the filter*

The decision is given after two operations on the signal $y(t)$ at the output of the triangular filter: the first is a sampling and the second is a comparison with a threshold. The numerical rate is $1/T$ bit/s. This means that there is one binary digit per T second(s).

PROBLEM.–

1) We suppose that $T = T_0$. Before sampling, the signal is of the form:

$$y(t) = y_1(t) + n'(t)$$

What is the expression of $y_1(t)$?

Also, what is the variance σ'^2 of the noise $n'(t)$? (During this computation we will use Parseval theorem, and we will admit that $n'(t)$ is a Gaussian-centered noise.)

2) We are always in the case where $T = T_0$. Give a representation of $r(t)$ and of $y_1(t)$ from the random suite $\{a_k\} = \{+1, +1, -1, +1, -1, -1\}$.

3) $T = T_0$. What is the best sampling time to decide of the values of a_k that gives the minimum error?

4) In this case determine the error minimal probability.

5) We decide to increase the rate T. What is the maximum of this rate that gives no I.I.?

SOLUTION.–

1) Without noise the signal is of the form:

$$r(t) = \sum_{k=-\infty}^{+\infty} a_k \delta(t - kT)$$

where $a_k = 1$ or $a_k = -1$ with the same associated probabilities 1/2. There is one emission of a_k per T second(s). If $T = T_0$, there is one emission of a_k per T_0 second(s).

$$x(t) = r(t) + n(t) = \sum_{k=-\infty}^{+\infty} a_k \delta(t - kT_0) + n(t)$$

$$y(t) = x(t) * h(t) = [\sum_{k=-\infty}^{+\infty} a_k \delta(t - kT_0)] * h(t) + [n(t) * h(t)]$$

It is:

$$\begin{cases} y(t) = y_1(t) + n'(t) \quad \text{with:} \\ y_1(t) = \sum_{k=-\infty}^{+\infty} a_k h(t - kT_0) \\ n'(t) = n(t) * h(t) \end{cases}$$

where $n(t)$ is a white Gaussian noise with a bilateral density power $N_0/2$ and $n'(t)$ is also a white noise of variance σ'^2.

Its density probability is then:

$$f(n') = \frac{1}{\sigma'\sqrt{2\pi}} e^{-n'^2/2\sigma'^2}$$

The autocorrelation function is:

$$P_{n'} = R_{n'n'}(0) = E[n'^2(t)] = \int_{-\infty}^{+\infty} n'^2(t) f(n') dn'$$

This gives:

$$P_{n'} = \int_{-\infty}^{+\infty} \frac{n'^2(t)}{\sigma'\sqrt{2\pi}} e^{-n'^2/2\sigma'^2} dn'$$

We make the transformation:

$$u = \frac{n'}{\sigma'\sqrt{2}}$$

And using:

$$\int_{-\infty}^{+\infty} u^2 e^{-u^2} du = \frac{\sqrt{\pi}}{2}$$

we have:

$$P_{n'} = \frac{2\sigma'^2}{\sqrt{\pi}} \int_{-\infty}^{+\infty} u^2 e^{-u^2} du = \sigma'^2$$

But from Parseval theorem:

$$P_{n'} = \frac{N_0}{2} \int_{-\infty}^{+\infty} |H(f)|^2 df = \frac{N_0}{2} \int_{-\infty}^{+\infty} |h(t)|^2 dt = \frac{2N_0}{3T_0}$$

and:

$$\boxed{\sigma'^2 = \frac{2N_0}{3T_0}}$$

2) The representations of $r(t)$ and of $y_1(t)$ with $\{a_k\} = \{+1,+1,-1,+1,-1,-1\}$ are (Figure 13.3):

$$r(t) = \sum_{k=-\infty}^{+\infty} a_k \delta(t - kT)$$

$$y_1(t) = \sum_{k=-\infty}^{+\infty} a_k h(t - kT_0)$$

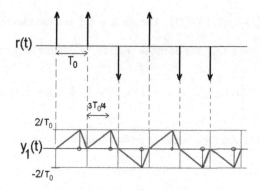

Figure 13.3. *Representations of* $r(t)$ *and* $y_1(t)$. *For a color version of the figure, see* www.iste.co.uk/jarry/communications.zip

3) We have to sample $y(t)$ and to compare these samples with a threshold. At the moments where $y_1(t)$ is maximum, we will have a minimum error. The sampling times will be:

$$kT_0 + \frac{3}{4}T_0$$

At these sampling times the levels are at a maximum $\pm 2/T_0$, then we choose a threshold of 0 volts (Figure 13.4).

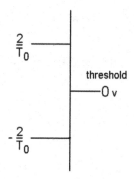

Figure 13.4. *The threshold*

4) We have $y(t) = y_1(t) + n'(t)$. The noise $n'(t)$ is a random signal.

As shown in Figure 13.5, the decision is given by the total error probability:

$$\mathscr{P}_e = p_1 \mathscr{P}_\varepsilon \left\{ 0 \le \varepsilon \le \frac{2}{T_0} \mid a_k = +1 \right\} + p_{-1} \mathscr{P}_\varepsilon \left\{ -\frac{2}{T_0} \le \varepsilon \le 0 \mid a_k = -1 \right\}$$

where $p_1 = p_{-1} = 1/2$ are the associated probabilities to $a_k = 1$ and $a_k = -1$.

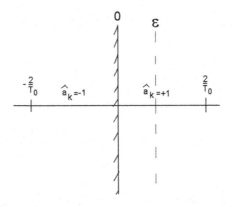

Figure 13.5. *The decision in the case of a Triangular Filter*

$$\mathscr{P}_\varepsilon \left\{ 0 \le \varepsilon \le \frac{2}{T_0} \mid a_k = +1 \right\} = \Pr ob \left[n'(t) \le -\frac{2}{T_0} \right] = \frac{1}{2\pi} \int_{-\infty}^{-2/T_0} \frac{1}{\sigma'} e^{-n'^2/2\sigma'^2} \, dn'$$

where n' is a symmetric Gaussian noise (Figure 13.6).

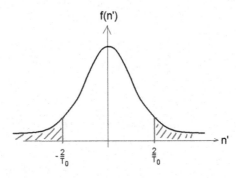

Figure 13.6. *Gaussian symmetric noise. For a color version of the figure, see www.iste.co.uk/jarry/communications.zip*

Then we also have:

$$\mathcal{P}_\varepsilon \left\{ 0 \le \varepsilon \le \frac{2}{T_0} \mid a_k = +1 \right\} = \mathrm{Prob}\left[n'(t) \ge \frac{2}{T_0} \right] = \frac{1}{2\pi} \int_{2/T_0}^{\infty} \frac{1}{\sigma'} e^{-n'^2/2\sigma'^2} dn'$$

This gives:

$$\mathcal{P}_\varepsilon \left\{ 0 \le \varepsilon \le \frac{2}{T_0} \mid a_k = +1 \right\} = \frac{1}{2}[1 - erf\sqrt{\frac{3}{N_0 T_0}}]$$

And also:

$$\mathcal{P}_\varepsilon \left\{ -\frac{2}{T_0} \le \varepsilon \le 0 \mid a_k = -1 \right\} = \frac{1}{2}[1 - erf\sqrt{\frac{3}{N_0 T_0}}]$$

And the total error probability will be:

$$\mathcal{P}_\varepsilon = \frac{1}{2}[1 - erf\sqrt{\frac{3}{N_0 T_0}}]$$

5) Now we have $T \ne T_0$ and we increase the rate of the numerical transmission. Then T becomes inferior to T_0. If the rate is very important (Figure 13.7), we can have the I.I. There is a contribution of the adjacent samples.

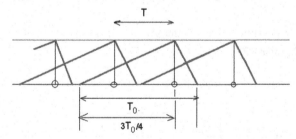

Figure 13.7. *Important rate with I.I. For a color version of the figure, see www.iste.co.uk/jarry/communications.zip*

As shown in Figure 13.8, there is no I.I. if we sample at the moments:

$$kT + \frac{3T_0}{4}$$

At this sampling time, we have no contributions of the adjacent samples.

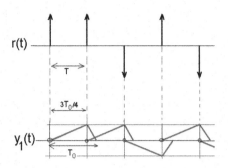

Figure 13.8. *Increase rate without I.I. For a color version of the figure, see www.iste.co.uk/jarry/communications.zip*

13.2. Detection of a digital signal with the presence of noise

PROBLEM.–

1) We consider the sampled and noisy signal $x(t)$ at the input of the R.R. of the form:

$$x(t) = \sum_k a_k s(t - kT) + n(t)$$

where $s(t)$ is the elementary pulse and $n(t)$ is considered as a white noise.

What are the possible values of the samples a_k in the cases of the three codes: NRZ, RZ and bipolar RZ?

2) The regeneration of the signal is made by using the system shown in Figure 13.9.

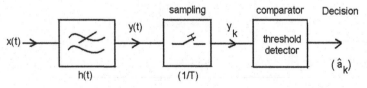

Figure 13.9. *Regeneration of the signal*

The input filter is an optimal receptor. What is its impulse response $h(t)$?

3) What, then, is the expression of $y(t)$ and of the samples y_k?

4) In the case of the NRZ code, the samples are compared to a threshold S as follows:

$$\begin{cases} if \ y_{k+1} \succ S & then \ a_k \ was \ 1 \quad (\hat{a}_k = 1) \\ if \ y_{k+1} \prec S & then \ a_k \ was -1 \quad (\hat{a}_k = -1) \end{cases}$$

What is the error probability $\mathcal{P}_e$ in the case of a Gaussian distribution of the noise of variance σ, with a threshold $S = 0$ and if the two states have the same probability 1/2?

SOLUTION.–

1) The possible values of the samples a_k:

– in the case of the NRZ code are $a_k = 0$ or 1;

– in the case of the RZ code are $a_k = 0$ or $(1,0)$;

– in the case of the bipolar RZ code are $a_k = 0$ or $(1,0)$ or $a_k = 0$ or $(-1,0)$.

2) The impulse response $h(t)$ of the input filter that is an optimal receptor is an adapted and returned filter:

$$h(t) = s(T-t)$$

3) And we have:

$$y(t) = x(t) * h(t)$$

or:

$$y(t) = x(t) * s(T-t)$$

By using the definition of $x(t)$:

$$y(t) = \sum_k a_k s(t-kT) * s(T-t) + n(t) * s(T-t)$$

or:

$$y(t) = \sum_k a_k [s(t) * \delta(t - kT)] * s(T - t) + n(t) * s(T - t)$$

$$y(t) = \sum_k a_k [s(t) * s(T - t)] * \delta(t - kT) + n(t) * s(T - t)$$

$$y(t) = \sum_k a_k u(t) * \delta(t - kT) + n(t) * s(T - t)$$

where:

$$u(t) = s(t) * s(T - t) = \int_{-\infty}^{+\infty} s(\tau) \cdot s(T - t + \tau) d\tau$$

But the quantity $u(t) * \delta(t - kT)$ is maximum for $t = T$:

$$\boxed{u(T) = \int_{-\infty}^{+\infty} s^2(\tau) d\tau = \int_0^T s^2(\tau) d\tau = E}$$

For the noise, we have:

$$\beta(t) = n(t) * s(T - t)$$

$$\beta(t) = \int_{-\infty}^{+\infty} n(\tau) \cdot s(T - t + \tau) d\tau$$

In particular:

$$\boxed{\beta(T) = \int_{-\infty}^{+\infty} n(\tau) \cdot s(\tau) d\tau = \int_0^T n(\tau) d\tau}$$

The expression of the samples is then:

$$\boxed{y((k+1)T) = y_{k+1} = a_k E + \beta((k+1)T)}$$

where we have replaced T by $(k+1)T$.

4) We compare y_{k+1} to the threshold $S = 0$. The decision is as shown in Figure 13.10.

$$\begin{cases} a_k = +1 & \text{if} \quad y_{k+1} = E + \beta \geq S = 0 \quad \Rightarrow \quad \beta \geq -E \\ a_k = -1 & \text{if} \quad y_{k+1} = -E + \beta \geq S = 0 \quad \Rightarrow \quad \beta \geq +E \end{cases}$$

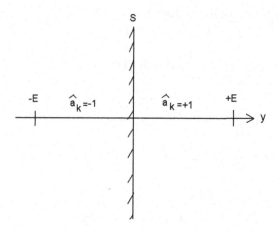

Figure 13.10. *The decision in the case of an input filter as an optimal receptor*

The total error probability with $S = 0$ is:

$$\mathcal{P}_e = p_1 \mathcal{P}_\beta \left\{ y_{k+1} \leq 0 \,\middle|\, a_k = +1 \right\} + p_{-1} \mathcal{P}_\beta \left\{ y_{k+1} \geq 0 \,\middle|\, a_k = -1 \right\}$$

where p_1 and p_{-1} are the weight functions.

If $p_1 = p_{-1} = 1/2$, the total error $\mathcal{P}_e$ can be written as:

$$\mathcal{P}_e = \frac{1}{2} \mathcal{P}_\beta \left\{ \beta \leq -E \,\middle|\, a_k = +1 \right\} + \frac{1}{2} \mathcal{P}_\beta \left\{ \beta \geq +E \,\middle|\, a_k = -1 \right\}$$

We have a Gaussian distribution:

$$\mathcal{P}_e = \frac{1}{4} \left(1 - erf \frac{E}{\sigma\sqrt{2}} \right) + \frac{1}{4} \left(1 - erf \frac{E}{\sigma\sqrt{2}} \right)$$

or:

$$\mathcal{P}_e = \frac{1}{2}\left(1 - erf\frac{E}{\sigma\sqrt{2}}\right)$$

where *erf* is the error function so that:

$$erf\, X = \frac{2}{\sqrt{\pi}}\int_X^\infty e^{-t^2}\, dt$$

and:

$$\mathcal{P}_\beta\left[\beta \geq B\right] = \frac{1}{\sqrt{2\pi}\sigma}\int_B^{+\infty} e^{-\xi^2/2\sigma^2}\, d\xi = \frac{1}{2}\left(1 - erf\frac{B}{\sigma\sqrt{2}}\right)$$

13.3. Digital transformation of an analogic signal

PROBLEM.–

1) We consider a coding system with a dynamic (maximum and minimum) $\pm V_{MAX}$ and with n bits of coding (including sign).

What is the quantification noise?

2) Give the ratio S/B_q.

3) We have a white noise of variance σ^2, what is the ratio now?

4) We impose a coding quality so that the ratio S/B_q is 30 dB with a dynamic of the input power $-40dB \leq P_i \leq 0dB$. What is the number of bits n we have to use?

SOLUTION.–

1) The quantification noise is:

$$B_q = \frac{1}{12}\left(\frac{2V_{MAX}}{2^n}\right)^2 = \frac{V_{MAX}^2}{3.4^n}$$

2) The maximum power of the signal is:

$$S = \frac{V_{MAX}^2}{2}$$

and the ratio is:

$$\frac{S}{B_q} = \frac{3.4^n}{2}$$

3) But if there is a white noise of variance σ^2, we have to do the change $S \to \sigma^2$:

$$\frac{S}{B_q} = 3.4^n \left(\frac{\sigma}{V_{MAX}} \right)^2$$

4) We have now in dB:

$$\left(\frac{S}{B_q} \right)_{dB} = 10 \log 3 + 20 \log 2^n + 20 \log \left(\frac{\sigma}{V_{MAX}} \right)$$

But the input power in dB is:

$$P_i = 20 \log \left(\frac{\sigma}{V_{MAX}} \right)$$

And we want and minimum input power of $-40\,dB$ with a ratio $(S/B)_{dB} = 30\,dB$.

We have with $\log 2 \approx 0.3$ and $\log 3 \approx 0.48$:

$$30 = 4.8 + 6n - 40$$

This gives:

$$n = \frac{70 - 4.8}{6} = 10.87$$

Then we must take:

$$n = 11$$

Bibliography

[CAR 87] CARLSON A.B., *Communication Systems – An Introduction to Signals and Noise in Electrical Communication*, 3rd edition, McGraw-Hill International Editions, 1987.

[COU 83] COUCH II L.W., *Digital and Analog Communication Systems*, Macmillan Publishing, New York, 1983.

[JAR 05] JARRY P., *Microwave Filters and Amplifiers*, Research Signpost, 2005.

[JAR 08] JARRY P., BENEAT J.N., *Advanced Design Techniques and Realizations of Microwave and RF Filters*, IEEE-Wiley, 2008.

[JAR 09] JARRY P., BENEAT J.N., *Design and Realizations of Miniaturized Fractals RF and Microwave Filters*, John Wiley & Sons, New York, 2009.

[JAR 12] JARRY P., BENEAT J.N., *Chapter on Miniaturized Microwave Fractal Filters – MMFF*, Wiley Encyclopedia of Electrical and Electronics Engineering, July 2012.

[JAR 14] JARRY P., BENEAT J.N., *RF and Microwave Electromagnetism*, ISTE Ltd, London and John Wiley & Sons, New York, 2014.

[JAR 15] JARRY P., BENEAT J.N, *Passive and Active RF-Microwave Circuits*, ISTE Press Ltd, London and Elsevier Ltd, Oxford, 2015.

[LAR 97] LARSON L.E. (ed.), *RF and Microwave Circuit Design for Wireless Communications*, Artech House, Boston, 1997.

[MAR 98] MARAL G., BOUSQUET M., *Satellite Communications Systems (Systems, Techniques and Technology)*, 3rd edition, John Wiley & Sons, 1998.

[SCH 87] SCHWARTZ M., *Information Transmission, Modulation, and Noise*, 3rd edition, McGraw-Hill International Editions, 1987.

[STR 82] STREMLER F.G., *Introduction to Communication Systems*, 2nd edition, Addison-Wesley, 1982.

[TAU 86] TAUB H., SCHILLING D.L., *Principles of Communication Systems*, 2nd edition, McGraw-Hill, 1986.

[ZIE 85] ZIEMER R.E., TRANTER W.H., *Principles of Communications, Systems, Modulation, and Noise*, 2nd edition, Houghton Mifflin Company, Boston, 1985.

Index

Printed in the United States
By Bookmasters